TIME OVERRIDE:
Prophetic Images From the Future Decoded

Dylan Clearfield

Copyright 2020 by Prism Thomas

Published by G. Stempien Publishing Company

All rights reserved

ISBN 978-0-930472-48-1

Photographs not the author's are courtesy of National Archives except where noted differently.

gfer

CONTENTS

Preface
Introduction
1. Beginnings
2. Original Shock
3. Windows of Prophecy
4. First Images
5. Vision by Faith or Logic
6. Signs of Disaster and Misery
7. Possible Explanations for the Images
8. Aliens and Concerts
9. 666 Identified
10. Codes in Living Glass
11. Demon and Friend
12. Absolute Proof
13. Findings

PREFACE

It is written in prophecies of the End Times that information will be revealed at a future period which could only be understood by the people who are living in that future period. We are that people, and in 2008 just such encrypted information was discovered, and in a most unexpected place – in the window panes of a historic courthouse.

A series of images were discovered that were transmitted onto a selected group of window panes but the images could only be seen after being processed by computer and scanned by a device possessing exceptionally powerful magnification ability. The nature and subject matter of the images was shocking. More than that, their very existence was inexplicable. How could such images even be?

Since 2008 much investigation has been carried out on the source of the images and the way in which they were created. In the process, several versions of this report were written that scientifically explained them. But they were explained only by expanding the laws of science to their limit, including quantum physics, plus inventing a special process.

While these explanations were viable to a degree they were still difficult to make fully convincing. And when it occurred that it was more rational and statistically reasonable to attribute the phenomena in question to the Will of the Almighty Creator (God) or Universal Intelligence than by explaining them by logical scientific formulae it was then realized that the most accurate version of the story was that the images are the product of Divine Revelation (Divine being howsoever a person chooses to define it).

For accuracy, what follows is the scientific explanation of this study: a building was struck by a particle-sized black hole filament and sustained an alteration to its subatomic structure. This caused the glass in the surviving window panes to become receptors of visual wavelength transmissions from the future which could be photographed only by quantum imaging. Quantum imaging is a very specialized form of photography which requires a black hole type of environment in which to operate. A coded warning was contained in these visual projections. The images themselves may have been copied and transferred to our realm from a duplicate history that is occurring in a parallel universe to our own. The apparent source of the message was Artificial Intelligence from our future which by sending it was either attempting to stop its own rise to power or to facilitate it. Thus is the science of it.

The investigation into the window images was finally concluded in 2020, a span of 12 years. The last key to deciphering their meaning was by recognizing that they were coded by visual cryptography. Visual cryptography uses images that are arranged in specific patterns and layers – often both – in which to present a coded message. Arranging these images into a "readable" format is how the code is deciphered. That's what will take place throughout this book, unveiling a warning of an Apocalyptic near future possibly from a Higher Intelligence of some nature.

One question must be addressed at the outset. The images that will be displayed will possess only minimum quality. None of them can be classed as visually sharp – but most of them are certainly identifiable as to content. The primary argument made is: if these images are the product of a higher Intelligence wouldn't they be of a better quality?

The answer to this is to refer back to prophecies of the past, many that were described in the Bible. The Dreams that Joseph interpreted were not easily deciphered. The Wheels in the sky observed by Ezekiel needed a great deal of consideration before they could be understood. Daniel prophesied in codes that required great Biblical scholars like William Miller to understand. Seldom are prophetical mysteries given in easily comprehended format.

Also, if all of the images that are about to be shown in this book were of the highest quality – wouldn't you suspect their source?

A final point about the Creator's power over time. Did He or the Universal Intelligence once stop the sun in motion and thereby cause quite a disruption in "normal" time? People who don't believe in miracles don't believe this biblical story. I choose to still believe in miracles, and science is used to create most of them.

What if the author of the images wasn't a Higher Consciousness but an alien species or even a representative of Artificial Intelligence from our distant future? While they may be able to transmit the prophetic images to us, they might not have been able to maintain the highest standard of visual quality in the process.

One additional matter. In all cases – the images are not illusions, fakes, or tricks of the eye. All of them are real!

Author February 29, 2020

INTRODUCTION

This is a book about a series of prophetic, Apocalyptic images that in coded form appeared in blank window panes in a historic, centuries old building in the United States. The images are coded in that to be understood they require a process of visual cryptography to be performed. They are not simple images that appeared in windows like the following in Clearwater, Florida.

Courtesy Bob Falcetti, Tampa Bay Times 1996

The prophetic images in this book began looking like this (all of them):

This was then subjected to computer processing (not enhancing or altering) in order to remove the overlaying darkness, which produced the following:

And the final processed image (not shown here) is then

transferred to a hyper-scanner for deep magnification to produce an image that has shape and substance which could be identified.

Later in the process this image is compared to several other images, related by their position, in order to produce an underlying coded message using visual cryptography.

God works in mysterious ways, it has often been noted. So too does quantum physics. Since quantum physics is theoretically the brainchild of God it would follow that they both work in the same mysterious ways. A simple logical syllogism.

I am a scientist but also a person of deep faith. However, I am as unwilling to extinguish my spiritual beliefs as I am to abandon my beliefs in science in order to explain both the existence and the meaning of the images highlighted in this book.

The images exist, they were produced by some method and if that method involves a scientific process it does not remove divine oversight from the process!

Remember, one of the greatest of scientists who followed the Christian faith, Sir Isaac Newton, predicted the end of the world based on biblical information. He calculated 2060 A.D. As being the year of Apocalypse. Ironically, that date coincides very closely with the time period suggested in the prophetic images revealed in this book.

BEGINNINGS

The story begins with this building:

It is here that the prophetic images were found. This building has a very long history and a very uncommon background. It was constructed in 1852 as a courthouse and was built in the the style of the King's College Chapel in Great Britain by the architect named James Renwick Jr. (pictured below)

Below is a drawing of the building at construction in 1852.

When Mr. Renwick designed this building, he followed the principles outlined in the book "True principles of Christian Architecture" written by Augustus Pugin. One of Renwick's closest advisors on the project was theologian Robert Dale Owen whose father Robert Owen was a famous theologian and utopianist.

Robert Owen founded the Christian utopian community of New Harmony, Indiana. His pseudo student, James Renwick, Jr., later went on to become a very famous architect, designing the old Smithsonian buildings in Washington D.C. and St. Patrick's Cathedral in New York. Obviously, there was a great deal of Christian influence in the construction of the secular courthouse in which the mysterious images were found.

And over the years, many tragedies befell this building and many direful life situations were played out here. It once served as hospital alternately for Union and then Confederate wounded. It had been an operational courthouse for decades where many souls had received sentences of death or long terms of misery in prison.

How many life and death legal cases were argued here? How many tales of suffering and terror were told here? How many damned souls walked through these aisles and may still be there today undergoing punishment and are displayed in the mysterious images in this book? Maybe mingled in with suffering souls of the future.

In addition, the building itself endured a great deal of physical damage in its time: surviving fires, military bombardment and, an enigmatic, biblically ferocious "windstorm" in 1858, blowing out over 400 of the building's window panes. The reason "windstorm" is and will be in quotes is because the atmospheric event that occurred here in April of 1858 was unaccountably mysterious.

The notices of the "storm" that were reported in the local paper, the *"Week Advertiser"* for May 22, 1858, described a storm that struck only this one building and blew outward from the inside most of its over 400 window panes. The rest of the town was undamaged. It seemed as if this specific building had been targeted for a singular type of attack.

There are many unusual structural features about this building. One of which is the the bell tower. It soars to an amazing height while housing a classic bronze bell fashioned by Paul Revere.

Not only is this bell tower unusual because of its soaring height but crowning it is a most suspicious weather vane. It is mounted so far above the ground that it is almost impossible to see which way the directionals are pointing or the markings on them. Not only that, it appears that only half of the usual compliment of directionals – which point out the directions the wind is blowing – are even attached to the vane.

This building passed through many renovations. The plans of various contractors who worked on the structure were studied by myself and it was discovered that the windows in which the prophetic images appear are among the few original windows left in the building. This can also be authenticated by examining the glass itself which displays the wavy appearance that was common to glass of the late 19th century.

Although the building began as a courthouse, it had filled many other roles in the early and mid 19th century as well as into the 21st century, often acting as a community center. It was the scene of various musical concerts, boxing matches, theater plays and even political events. This will be important when identifying the images in the upcoming windows because they seem to originate from many of the events just mentioned as if movies of the past and parallel time are being replayed on the window glass using it as tiny movie screens, mixing past and future together.

At one time, a portion of the building was even used as a firehouse. Ironically, it was abandoned as such after a fire destroyed most of that section.

This building is now off limits and strictly closed to the public. This is fact, even though the people in control of the building claim that tours are available and that it is open to the public.

They have even posted an online, fully working website for the building in question, showing it to be a normal office building doing business as usual. It isn't. This is a deception – misinformation. It is meant to devalue any factual findings and to cause the reader to believe that the building displaying the prophetic images is actually only an ordinary structure that shows no evidence of change.

It isn't known who wants to keep this discovery secret. Or why. Just that someone does. Probably to keep sight seers away, as in most cases like this one.

But there exists another very critical reason why this location is protected from visitors. In the first decade of the 2000's, the building engineer took the major, risky step of ordering the first floor of this much used official county governmental building closed and **EVACUATED IMMEDIATELTY** upon his shocking findings. This is a fact!

The upper floor was not in use at the time.

The evacuation was made so quickly that the personnel working there were instantly removed to another building, leaving all of their materials behind, even their own personal items. They were never allowed to return to their old offices or to retrieve their "infected" belongings.

Upon peering through the windows and into the interior of the building during this investigation a most eerie sight was on view. Everything seems to be in a a state of suspended animation – frozen still not just deserted!

The building's chief engineer was interviewed and was directly asked about the sudden evacuation of the courthouse. His only answer was that, while his crew was engaged in a routine renovation of the building, a <u>critically hazardous</u> condition was discovered which warranted immediate removal of all personnel. He would not specify what this critically hazardous condition was but made the strange allusion that it was related to exposure to a lethal temporal environment! He did volunteer that the building was structurally very sound and that the danger was also not from radiation, asbestos or any common environmental hazard.

Or maybe a cosmic shock had been discharged into this structure long ago by forces whose power is only now being felt.

ORIGINAL SHOCK

The investigation started with this image:

A marathon day of picture taking of some of the most beautiful Antebellum architecture in the area was just being completed when I was suddenly called to take one more picture - the one above. I had already bypassed the building which contained this bank of window panes.

But something – call it divine inspiration – forced me to sharply turn around and take this one quick picture before continuing immediately onward. It was as stark as that: take this picture. And it was with this picture that the mystery of the prophetic images began.

Being a field archaeologist, architecture is among one of my many interests and I had taken hundreds of pictures that day with a newly purchased, most advanced digital camera of the time. This was on September 9, 2008. The data from digital cameras must be downloaded onto a computer to be processed and that is when the results of your photographic work is seen. For reasons unknown, during photo processing I was highly attracted to the picture that was highlighted above, the one that I had been summoned to take.

Following this latest inspiration, I enlarged and magnified a group of window panes, not sure what I was looking for. That's when I spotted it! An odd shape that didn't seem to belong there.

With continued adjustment of the image, the shape was brought into sharper focus – without causing any change to the original data – and then the following image was produced.

Please don't be alarmed by the terrible quality of the above image or the next series of images because they are not representative of the others that follow. In fact, these do not even belong to the group of prophetic images to be examined later.

There **IS** something in the above window. It's almost impossible to see, but there is an image behind the window. It is sitting sideways behind those six panes with one arm outstretched onto its knee.

I first suspected that it might be a person seated there but quickly discounted this because the height of that section of window spans 12 feet. This would have to be an unbelievably tall person.

Another reason it couldn't have been a person was because this building was closed; it is the same building that was introduced at the beginning of this book.

No one should've been in this building; and it was certain that no one was in the building.

Maybe the image was a huge cardboard advertisement that had been propped behind the window bank. But, for reasons that will soon be apparent, that idea had to be discounted.

Additional adjustments were made on the image in order to delineate the shape of whatever was behind those windows.

Changing it to a matte like finish made the picture grainier but the shape inside of it a little easier to see (as in the images below).

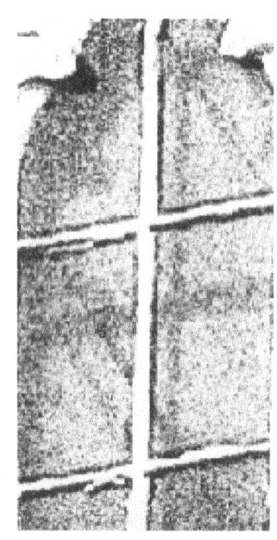

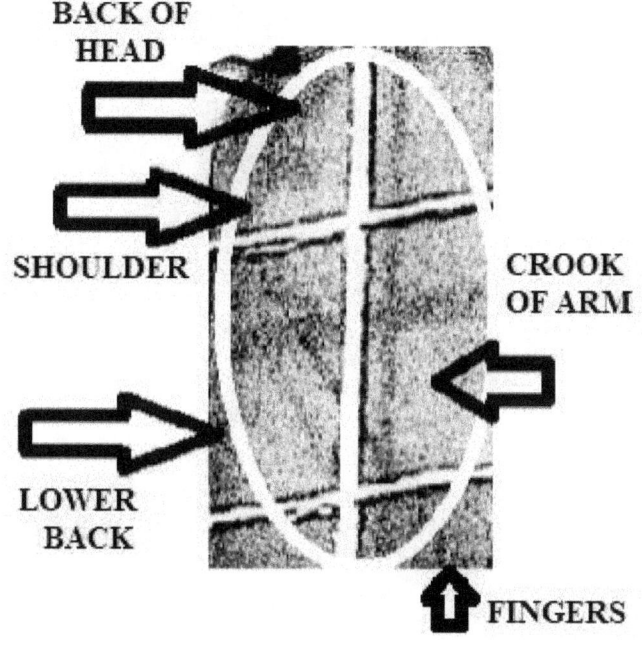

The next picture is enhanced in an attempt to show the horn-like features on the creature's head (arrow) and what looked like webbed fingers (arrow). A face can plainly be seen. Nothing has been added or deleted. Background was darkened to reveal image.

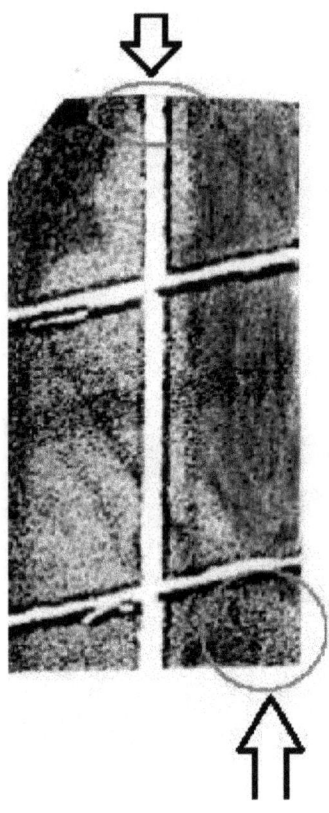

This looks like what Satan is usually depicted as. Below is an enlargement.

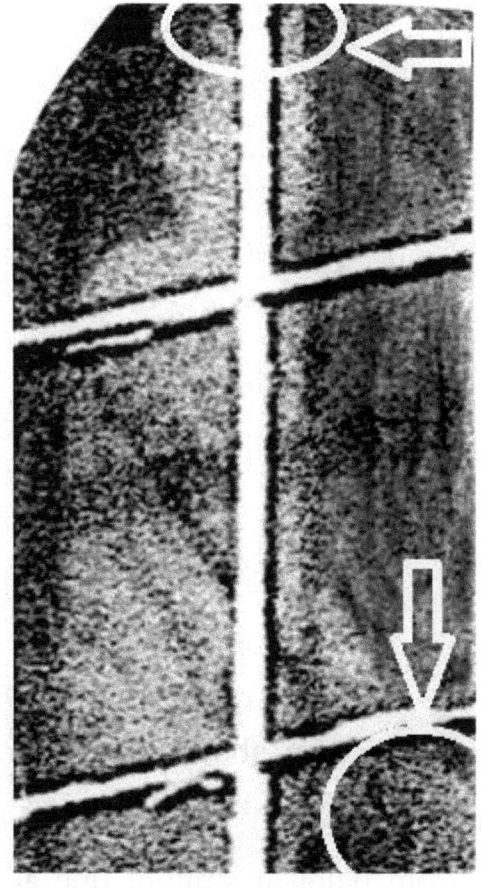

But is there really something tangible there, something with substance like a body and not merely a flat cardboard cutout? The next picture answers that query.

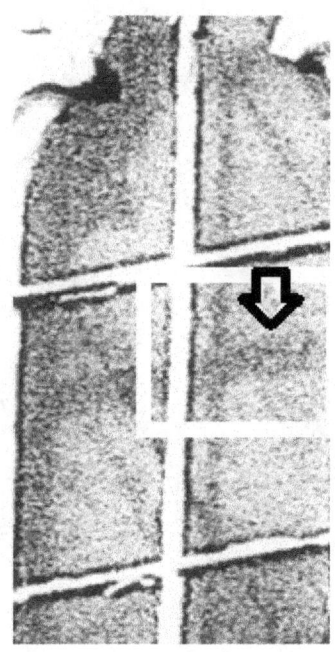

Within the white square above there is a tiny black arrow pointing to a thick, dark line. At the spot where the arrow points the line bends slightly from being a straight line and forms a sudden angle. The line is a shadow made by a muntin (divider between panes).

This is of critical importance because it proves that something of physical substance must be sitting behind the window to cause the shadow falling across the muntin (divider) to change from a straight line and then into an angle. An arm? Whose arm? Satan's? A minor demon's? And was it his silent voice that called on me to take that hasty photograph of the building?

But the silent – mentally heard – voice that inspired the picture taking felt benevolent. It's purpose was a good one, not satanic. Meant to expose Satan rather than glorify him.

Ironically, the image just examined is not among the prophetic ones. It did not require visual cryptography to decode it or explain it. It was simply there. A lure?

Even though the image seems satanic, it doesn't appear that the motive behind its existence is evil. Because, as noted, this particular image didn't require sophisticated visual cryptography to decode. Even though it is of exceptionally poor quality its appearance was visible and by presenting a mystery it provided the impetus that led to the discovery of the other more guarded prophetic images which otherwise might have gone unseen because they would not have been searched for.

WINDOWS OF PROPHECY

This chapter is going to provide a basic description of the physical route that was taken in the discovery of the images and will give an idea as to their positioning in the building itself.

Amid all of the hundreds of panes and banks of windows in this building, there are only 2 banks of windows and only 4 actual panes from which the images emerged. And they are placed on the same side of the building – facing West – and are separated by only one other bank of windows. They are all among the handful of original panes that survived the "windstorm" of 1858!!

And in those special banks of windows, the only panes that revealed prophetic images are shown filled in white below.

Yet, a search had been made of pictures taken of every window pane in the building, and the only positive results appeared within a few feet of each other. Over 400 other panes!

Other strange images were captured in windows in other locations throughout the building but, like the satanic image, they were not of the coded variety, not requiring visual cryptography to raise from obscurity. These proved to be simple curiosities.

As a matter of thoroughness the panes which produced prophetic images are shown below in larger size. Each group of 4 panes that bear images will be referred to as a BANK. Thus, the illustrations are identified as bank 1 and bank 2. In bank 1 the specific locations of the various images are designated by either a circle, oval or rectangle. In bank 2 by ovals.

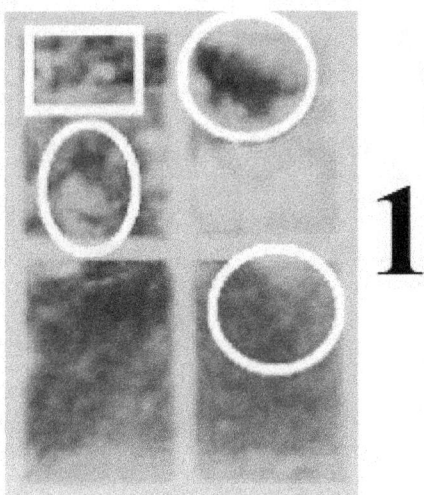

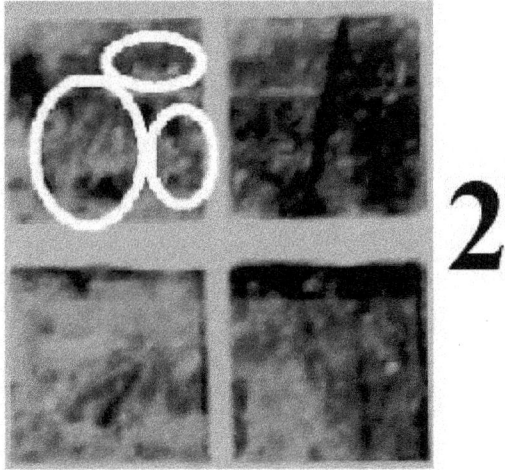

FIRST IMAGES

Now that the lure had been dangled, the search for any additional strange images in the windows of the mysterious courthouse was begun. The building was revisited and hundreds of exterior digital photographs of the building were taken, focusing on the windows.

Pictures were taken of every pane in the structure, amounting to over 400. The interior was off limits to everyone, including a fellow scientist who had been performing a study of the marvelous rafter work (above) on the upper floor.

The above picture of the artistically designed rafters is decades old and was obtained from the historical society. It should be remembered that this building was designed following basic Christian architectural principles as provided for in the work of A. G. Pugin as previously noted. There is a great deal of spiritual influence inculcated in this building and this may be why it was chosen to preserve and provide sanctuary for prophetic messages for future generations.

If this be so, then how to account for the biblical "windstorm" that struck this building in 1858 – only 6 years after its construction – which destroyed so many of its original windows? Maybe the purpose was to prepare and protect the important surviving window panes!

This requires explanation and a certain amount of suspension of belief – or adoption of belief, depending on one's point of view.

If a Supreme Intelligence wishes to perform what might be called a miracle, might not this Entity do so by using the forces of Nature and Science for His own purpose even if it involves breaching the so-called Laws of that scientific domain? A miracle by any other word is still a miracle.

Some people argue that the Creator cannot break his own Laws of Universal design, but if He is the Creator of them why can't he?

At any rate, the proposition being made now is that the great "windstorm" that struck the building in question, upon doing so, imbued the special, prophetic windows with a type of magnetic coating or resistance which both preserves and obscures the images that reside within the glass?

This "coating" can be penetrated by the use of digital processing and scanning upon a hyper powerful system to then reveal the hidden images beneath. Additionally, the images are arranged in such a way that they must be decoded using visual cryptography if they are to be understood.

There is basic, verifiable evidence to suggest that the powerful "windstorm" assault of 1858 has had a drastic impact upon this building, and specifically its glassy surfaces.

The destroyed window panes were replaced after the so-called "windstorm" and it was not long afterward that very bizarre phenomena began to plague this building; and it was recorded.

The first sign of odd events taking place was when unusual visual anomalies began to appear in the window glass of the first floor courtroom in 1887. An unearthly brightness seemed to radiate from the windows which had not been seen prior to the storm –

neither the newly replaced glass, nor the old glass. The presiding judge complained of the eerie and annoying radiance and ordered all of the westward – street side – facing windows be painted dark green because of the unnerving glare.

This painting was performed by a man named R.L. Jefferson that same year. Eventually this paint was scraped off but the oddity of having windows painted over seems glaring in itself. What had happened to the glass to have made it so unbearably radiant?

The arrows point to the areas of green that hadn't been fully scraped away.

This gives a strong indication that there is something very peculiar about the environment surrounding this building. Taking into account the just mentioned theory of how the prophetic images became ingrained into the window surfaces, the process for how they can be digitally risen into view comes next, starting with the simple blank window pane (bank **2**) that offered an earlier example. This is the upper left pane.

And by following process outlined earlier, that is how, this:

Became:

And was resolved into this:

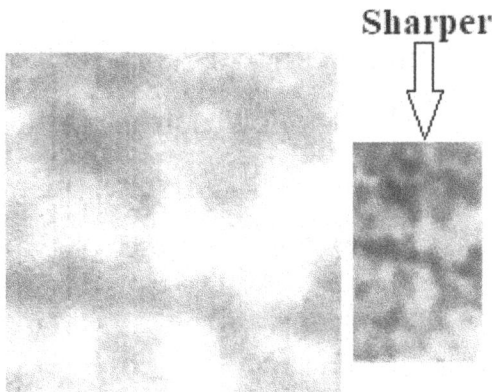

But those were only the initial stages. The above image in itself tells us nothing. It still isn't even that clear. It must be seen in relation to other images in order to acquire some form of message – other images that are right next to it in the same bank. That's the foundation of visual cryptography. First realizing the existence of the other images that are placed in relation to one another, either by being stacked in layers or hidden behind different visual quality, and then understanding their meaning when considered together.

We will begin with the above image. This represents a great deal more than a man who is wearing a white "polo" shirt and apparently sitting in the midst of a large crowd. He is leaning forward with his forearm on a railing and seemingly watching some event taking place before and below him.

Here is the subject image enlarged:

Smaller but sharper:

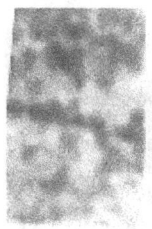

Within the rectangle is the man – head slightly turned – who is leaning over a railing and wearing a white "polo" shirt. While it isn't sharp or clear, it certainly is decipherable; even the style of his haircut could be determined. (Better view in smaller version).

There are other features in this frame that haven't yet been fully resolved into sight. Upon further magnification of the other portion of the image – the spot to our left of the man in the "polo" shirt – another human like being appears. He is in the circle in the picture below.

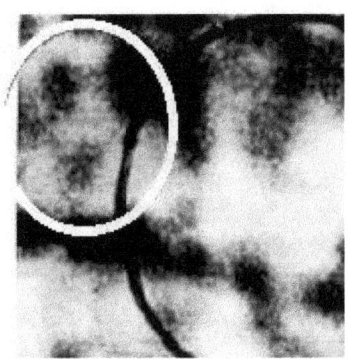

Shown is a balding man with a Van Dyke beard. Next to him, slightly more out of focus now, is the man in the "polo" shirt. Both seem to be in the midst of a crowd. Later it will be revealed why this is completely appropriate and important.

This is how visual cryptography is used by the encoder. The image of the balding man, while placed directly next to the man in the "polo" shirt, could only be seen if a much higher magnification was used on only that portion of the entire image in which this 2nd person is found. Otherwise, he would've remained an indefinable blur. Placing these two people side by side is building a coded message whose meaning still needs deciphering. Thus, more examination of the entire image is needed.

AND, there is a 3rd person in this image as yet unseen. But his appearance will present a mystery of amazing depth which will be examined later.

At this point, in our possession are at least 2 relatively clear images of 2 men which were raised out of a photograph of a bank of blank, deeply dark window panes. Images which shouldn't be there by ordinary standards.

How many other images were hidden under the outer darkness of the other windows? All of the windows when photographed presented blank surfaces without any hint of anything unusual about them. It required processing by computer to raise the first 2 images into view followed by detailed, precise scanning to magnify them and sharpen their clarity. Would there be any other images?

Thus began an exhaustive and painstaking process of examining the more than 200 remaining photographs of this building which had yet to be digitally developed.

Could the original image have been an anomaly? A photographic aberration that wouldn't be duplicated? A rare mixture of light, color and shadow, producing figures which weren't really there? A fluke?

Next, photographs of more than a hundred window panes

were investigated which gave negative results. But finally another positive result was secured. Another window pane was found that held an image.

But why did it take until 2008 for anyone to discover the images? Surely other people have taken photographs of this historic building since 1858. This problem was briefly mentioned earlier, but a more extensive explanation is in order.

There are a number of answers to this question as to why it took until modern times for these images to be found. One answer is because digital cameras must be used to perform quantum imaging of this type and they are a relatively recent development, compared to the full history of photography.

Also, computers must be used to process the images. It also isn't until relatively recently – as of the original writing (2008) – that computers have been used for this purpose.

Another necessary item is what is called a "hyper-scanner" which has the ability to magnify any images without losing resolution to a degree of 1700%. I had possession of such a device which is commonly not available outside of the military or highly restricted scientific labs. This scanner allowed the images seen in this book to be captured.

And, remember, as note earlier – according to biblical thought – certain information about the End Times was kept from being revealed until those times were near.

Another great mystery remains and must be addressed now. Not everyone can see these revealed images even after processing. How can that be accounted for?

VISION BY FAITH OR LOGIC

Before examining the many other images awaiting investigation, this disclaimer must be presented.

I was shocked one day when a person proofreading this book told me that she could not see any of the images. What! They're what this report is about. Was she joking with me? No, she wasn't. She could not see any of the images.

I asked another reader if she could see the images. Thankfully, she said that she could clearly see them. All of them. How was this disparity of vision possible?

One of the most confounding mysteries presented by the images in this book is that some people can clearly see them while others cannot see them at all. This enigma had to be confronted before this final version of the report was published.

An explanation was due to the people who read this work and honestly could not see any of the images. Either supply them with an explanation or allow them to believe that the entire investigation was a fraud. I am a scientist, not a purveyor of snake oil.

The first explanation that came to mind was that it was

possible that some people could not see the images for the same reason that color blind people cannot see certain images that are shown to them in eye exams used for testing color blindness. But this had to be rejected when it was learned that both color blind and non color blind people could see the images in this book.

Then a type of answer finally presented itself on February 27, 2015, approximately 8 years after the images were first discovered. The partial answer was found in an event that overwhelmed the internet across the world. It involves the mystery of the bridesmaid's party dress.

A woman in Scotland purchased a dress to wear at a friend's wedding. Without plan or warning, the internet was set aflame by discussions concerning a question about this dress over which many millions of people had a viewpoint. What was the true color of the dress that the woman bought to attend her friend's wedding in Scotland?

If you do not recall this famous dress incident, it centers around a single piece of attire that appeared in two different colors to the various viewers of it. It was one dress, but some of the viewers saw it as a golden color and the other viewers saw it as blue.

At the time, I performed a scientific experiment on the photographs of this dress. I applied a reverse color filter to the gold colored dress and when in reverse - it was blue. When I reversed the color of the blue dress - it changed to gold.

Does this imply that some people see reality in reverse? Which people though? Reverse of what? What is the true color of the dress? Maybe it's neither. It was difficult to comprehend how someone else looking at the exact same object could see it in a completely different way. It's as if you saw a stop sign painted gold and the person beside you saw it painted blue. You couldn't both be right – could you?

In the above image: is the one on the left gold or blue? Is the one on the right gold or blue? I see gold on the left and blue on the right. What about you?

In the case of the mystery dress, both views of the dress were right. This dress was being seen in the color blue and in the color gold at the same time by different people.

Scientific explanations then began to arise. Most of the

"scientific" views classed the phenomenon as a visual anomaly rather than a true distortion of reality. Those who argued this believed that some people saw the dress as blue because the number of color sensitive cones in their eyes were different from other peoples'. Or people who were more fatigued than others saw a different colored dress due to the fatigue factor. There were not many arguments made that simply stated that sometimes reality is truly different for different people.

None of the "scientific" answers were usable because none of them could account for all of the situations. If fatigue was a factor in the selection of color made by the tired eyes or mind, the opposite color should be seen when the person was no longer tired. This was not the case.

And as to the suggestion of different types of rods causing the viewer to see different colors? This couldn't be proved either. Neither could any of the other explanations that were attempted.

In the case of so-called spiritual images in this book might it be a matter of faith whether or not a person can see them? Highly doubtful. Take as a prime example the miracle of Fatima. Hundreds of people witnessed to the events of that day. Yet hundreds more claimed to have seen nothing out of the ordinary? Were the people who did not see the miraculous occurrences non-believers? Not by their accounts. Many of them were as pious in belief as those who'd witnessed the events.

The simplest and truest answer seems to be that people simply see reality differently at various levels and at various times.

What was the true color of the dress? Can you see the images in this book or not? Either way, the data is still fascinating. Keep reading! You in fact may notice something that no one else has. This is not an empty remark either. Some of these images have revealed sights to some people that others cannot see – including items that I myself have not noticed. The images are alive.

SIGNS OF DISASTER AND MISERY

Here begins the revelation of the prophetic images which directly demonstrate signs of catastrophe. The images that follow were retrieved from the top left pane of the grouping of windows below from bank 1.

It must take a certain amount of faith to look for images in a totally dark window pane (as in above)! And, in this case, the early reward was found in the next level of illumination which added a little detail.

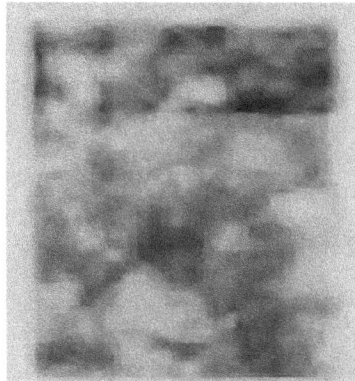

    A mere hodge-podge of blurry blotches? Or a form of visual cryptography where the hidden images are overlaid one upon the other and produced in different strengths of clarity, requiring different strengths of magnification for different portions of this single pane? The second answer would be the correct one because that is exactly what this blurry mass truly is.

    The first images of disaster will be removed from the above pane, magnified and clarified. The initial group to be studied are a pair of people in hazmat suits that are inside the square.

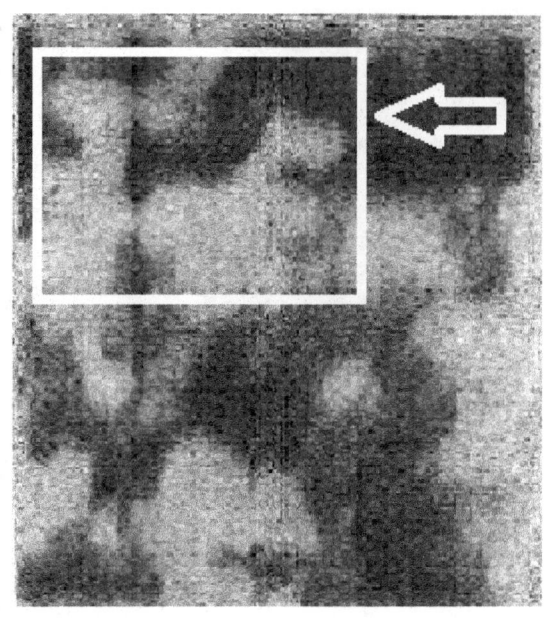

It will resolve into this:

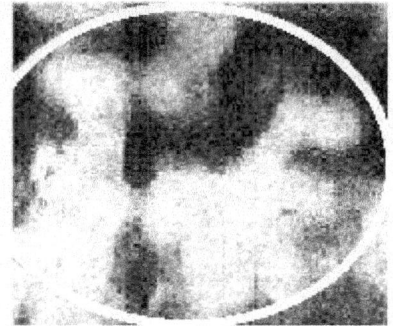

The enlargement below will exhibit more clearly an injury victim being carried over the shoulders of one of the rescuers. What

50

is being shown are two people (facing opposite directions) wearing hazard suits – or hazmat suits – one of whom is carrying what appears to be a sick or injured person over his right shoulder. The person in the suit carrying the afflicted person is on the right side of the picture. The arrow points down toward the victim and is directed straight toward his upturned face, (features not visible), head dangling far back.

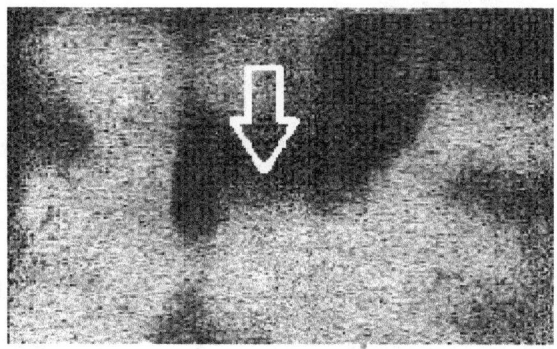

Below is an example of a more modern form of hazmat suit for comparison purposes. They hardly differ at all.

Compare them side by side below: modern with the End

Times image.

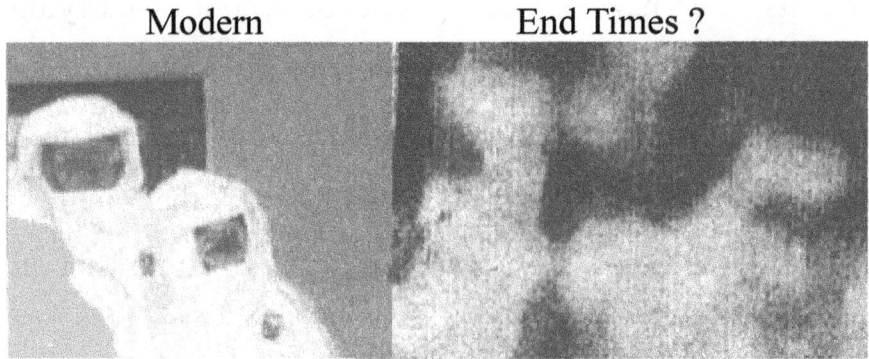

This is the first real indication that something is seriously wrong. People in hazmat suits are carrying victims of some form of catastrophe for treatment – or maybe burial.

Is this a view from the midst of an ongoing apocalypse? There isn't enough information in this one group of images to make that claim yet.

But there is much more in this individual PANE – a single frame – of multiple images. The next image to be examined appears below and slightly to the right in the original segment of the people in hazard suits. The arrow points to it.

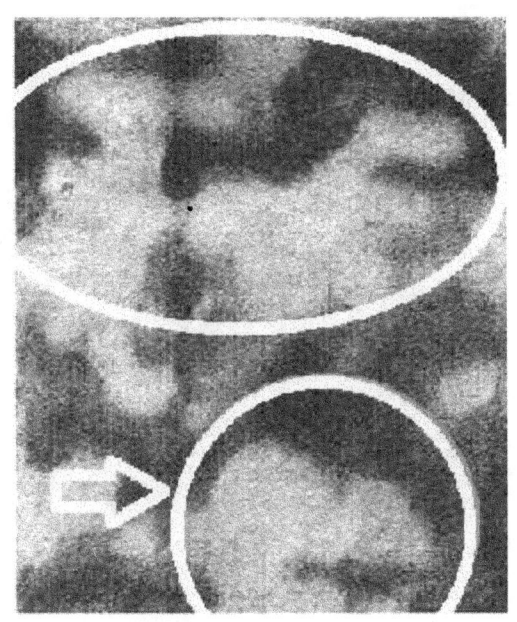

The image is a side view of a female who appears to weeping into a hand raised to her face.

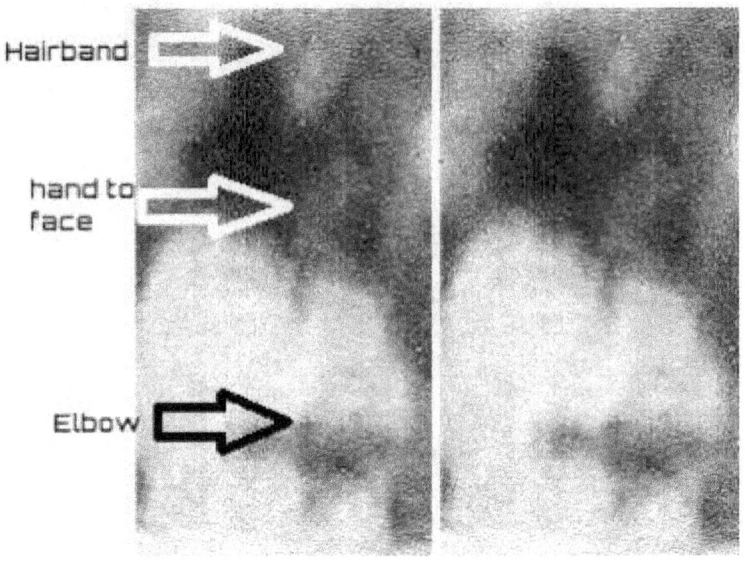

Where is this woman? Can a location be pinpointed? Judging from the chaotic surroundings, she too is in the midst of apocalyptic destruction. But she is not necessarily in the same location as the people in the hazard suits. Here is where detail will help. Examine the individuals to the woman's immediate left. Can you determine what they are doing?

The primary figure, call her the "penitent," is praying with the two people on her left. One of the people seems to be wearing a nun's habit common to the Catholic Church and the other person simply has head bowed and arms outstretched to the front with hands pressed prayerfully together.

The nun's head is also deeply bowed, so much so that her face is mostly hidden in its downward pointed direction. This is just one form of detail. Look closely and you can see her hands pressed to her eyes too, crying.

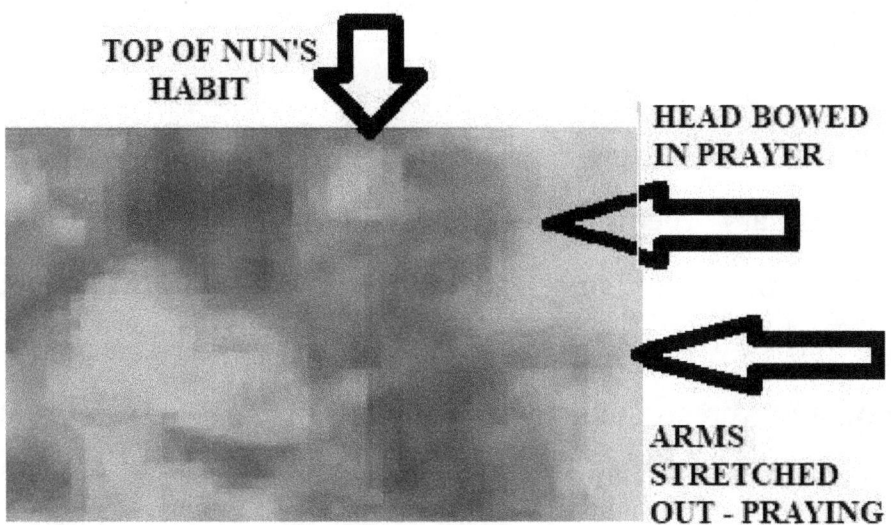

Another form of important detail is in precise attention to

environmental surroundings, including clothing. Detail in an image is vital when confirming the reality of a scene. Defining very small details in settings that were not easily visible to the naked eye greatly adds to the evidence and increases authenticity of the image. This is a major argument against those who claim that all of these images are simply a product of **pareidolia. Pareidolia** is an optical phenomenon in which the visual process of the eye and brain combine to form sensible patterns and images from what in reality are shapeless unconnected forms. That is the foremost explanation skeptics use to explain these inexplicable pictures.

But it is difficult to maintain this argument after viewing the next images. These were raised from the robe like attire being worn by the primary "penitent." This detail is the result of hyper-scanning being applied to a segment of her clothing. It was possible to raise into view a stitch in the sleeve of the woman's garment and a piece of stray thread that is hanging loose. This greatly helps affirm the physicality of and gives reality to the image. This is not a trick of the eye.

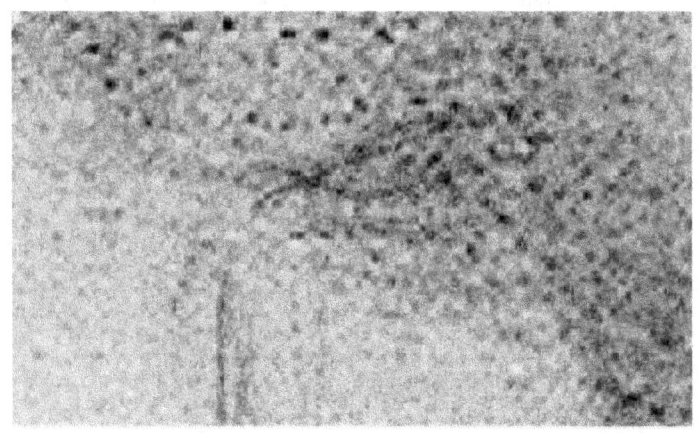

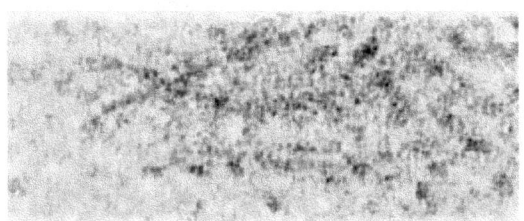

Does the above not look like a hastily sewn tear in a cheaply made garment with loose thread dangling free? What this implies is a world of poverty and misery potentially ruled by a madman dictator. Or an antichrist.

Is it possible to locate where this group of penitents is? One can assume it may be a church or some other sacred location where people might gather in prayer. But can anything more definite be produced to verify this? Examining a section of the frame from window pane 1, which contains the people in hazard suits and the praying penitents, there is a figure which may give an identity to the location of the setting.

The arrow is pointing toward what may be a very important personage.

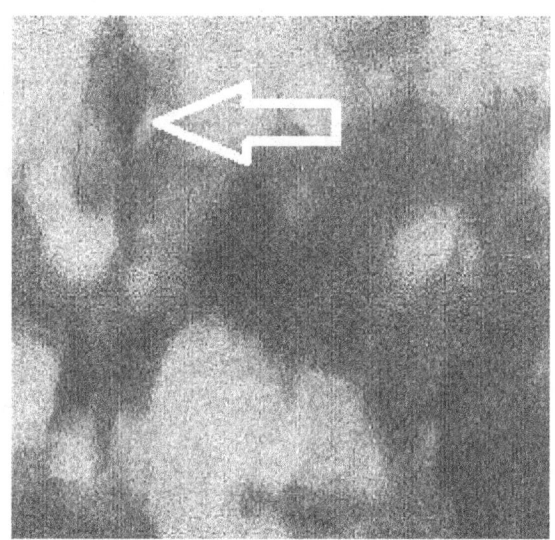

The arrow is pointed toward the mitred head of a man who is standing almost back to back with the primary penitent in the scene. This is the type of headgear this is traditionally worn by a bishop in the Orthodox Church which simultaneously symbolizes both the royal crown of Christ and his crown of thorns (seen below).

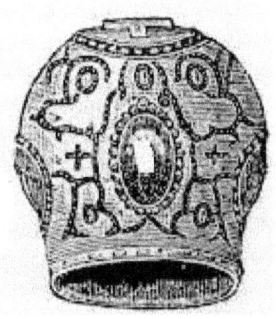

Mitre

Attention was drawn not only to the mitre that the person is wearing, but also to the wide collar around the neck which is called an amice.

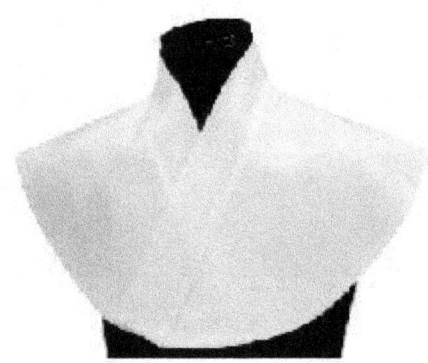

Amice

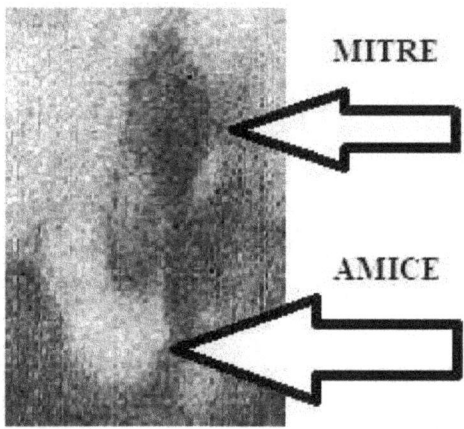

Admittedly, this image is of extremely limited visual quality – but there does seem to be enough detail available to reasonably speculate about the identity of this person, if person it is.

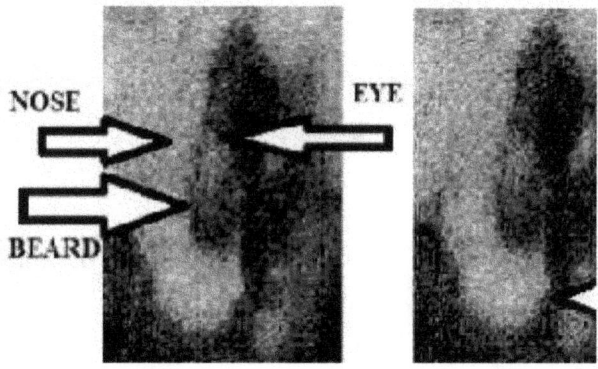

As speculated, this may be a bishop of the Orthodox Church. These contents seem to be primary elements that would be found in the End Times. The dead and the sick and injured being carried to help or burial by people in hazmat suits. Citizens gathered in prayer at what may be a church where a bishop of the Orthodox Church is in attendance along with a nun of uncertain denomination.

But can all of the images be explained in what may seem a more rational, realistic way? Every potential realistic explanation that this author believes can be conceived will be examined next.

POSSIBLE EXPLANATIONS FOR THE IMAGES

Isn't it obvious that the images are only reflections of objects in the surrounding environment on the glass? This objection is the most consistently put forward.

There are several reasons why the images are not reflections. 1) It was made certain there were not any reflections on the glass when pictures were taken. 2) There wasn't anything in the area that matched or could have made the images that appeared on the photos. 3) The images were proved to be in motion. 4) The images were in layers; when a top layer of images was processed from a photograph on the computer this often revealed other images below it. 5) The images were far too small to reflect any "normal" sized object in the environment. 6) No one or no object inside the building cast a reflection on the window or were photographed through the windows.

Identifying these images as being either projections transmitted from another reality or parallel time or by the grace of Universal Consciousness seemed the most likely to be true, despite how difficult to accept this may seem. All other logical explanations and interpretations for them were attempted but had to be rejected.

Beginning with the makeup of the windows themselves: one suggestion was that there is an anomaly in the physical structure of the glass which accounts for the images appearing when subjected to digital photography and processing by computer.

This was ruled out quickly. Having worked as an industrial engineer for one of the largest window and door producing firms in the world, I have an intimate knowledge of glass and its eccentricities. None of these windows have qualities that lend themselves to the creation of images. They were examined from very close range and by several other technicians as well.

Could the glass of the windows have acted as natural photographic plates? In early photography the photographic plates on which images were captured were made of glass. It was suggested that the windows of this building somehow act as photographic plates and retain images like any other photographic negative. The intense sunlight would have burned the images into the glass. This seemed like a very good hypothesis.

But even this idea had to be rejected. If the glass of the windows had acted as photographic plates the images captured on them would be permanent and the same pictures could be transferred from them at any time – like any other negative.

This isn't the case. When photographs of the same windows were taken later under the same conditions the same images no longer appeared after processing. And, still photographic images cannot show motion like these do.

Was the camera defective? All photographs were taken by the same camera, a completely new digital model just removed from the box. There had not been any pictures of people, animals, or anything else taken beforehand so even transference of image or double exposure can also be ruled out. The camera was in perfect condition and there were no defects in it, or in the computer that processed the images, or in the scanner that magnified them or in the software that was used. There were no technical defects of any type.

Is it possible that the images are just tricks of the eye? One of the most common arguments is that which claims they are in reality formless features from which the mind automatically creates recognizable figures. This is a condition mentioned earlier, known as **pareidolia.**

However, this explanation can be discounted since the images are clearly more than formless features and they will shortly be proved to be in <u>motion</u>. Of critical importance is that the figures are all in proper ratio. The subjects in the images have all body parts where they should be and in the length and proportion that would be expected. The figures are all anatomically and otherwise physically accurate which would be unlikely in all cases of pareidolia. Again, even though the images may be products of other times and other dimensions and are in many cases of very poor quality they retain their ratio integrity. Statistically this is too improbable to be considered coincidence or simple pareidolia.

When observing all of the source windows there isn't anything on the glass and there isn't anything behind the glass. So the question to ask next is: do these images somehow appear within the glass itself? Is there an ongoing world within the glass?

The answer is, no. The only function that the glass serves is as a form of screen upon which images are projected from another source possibly through the agency of the effects of a particle-sized black hole filament, though the actual agency of transmission is still unidentified.

The suggestion has been made that if photographs were taken of enough windows in random tests similar images to the ones being presented here would ultimately be found. It's a common occurrence, only an unnoticed one.

I considered that possibility, too. And I experimented with it to the point of exhaustion, taking hundreds upon hundreds of photographs of random windows in search of images like those that appear in these pages. None were found!

An intriguing potential explanation for the images is that they are thought forms. Can the images be thoughts and images that were in the minds of people who occupied or visited this building which somehow became ingrained into the very glass of the windows?

If so, this in itself would be an extraordinary phenomenon. What if someone who worked in this building spent long hours worrying about her pet cat at home? Could that image have become "burned" into one of the window panes? Below is a genuine window pane image.

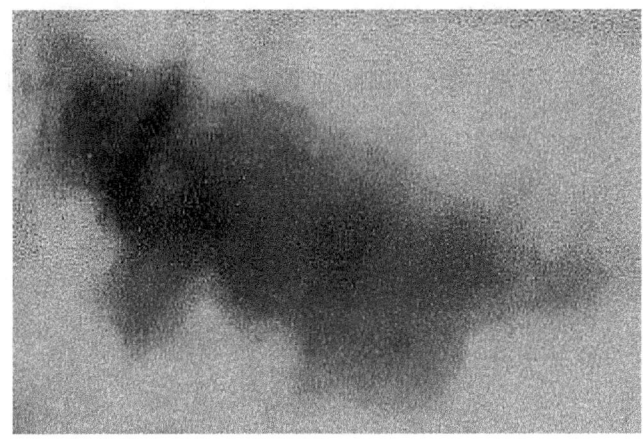

Can this be anything BUT a black cat?

But if these images might be produced by the thoughts in people's mind, would some of the images be more like nightmares?

ALIENS AND CONCERTS

What constitutes a nightmare image as mentioned earlier? Monster-like creatures or maybe settings that seem so outrageous and beyond normal appearance that their basic feature is a warped form of sense and reality? If that is so, the next group of images answers to that description. These too were found in the mysterious window images.

The bizarre images that follow were contained in the below bank of window panes. This is a from Bank 2 (first stage of lightening).

Below is a larger view of the top 2 panes (magnified).

The "nightmare scene" is within this selected pane:

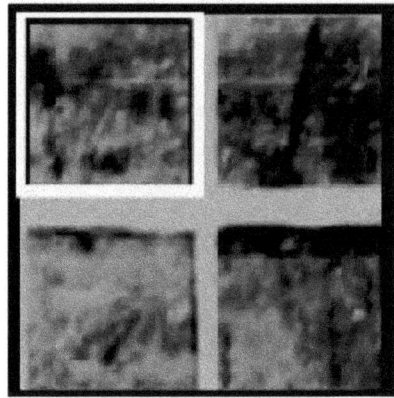

Recall, this is Bank 2 of images and only the top pane to the left revealed prophetic scenes. And they are surprising.

A magnified view of the target pane is next. Arrows point to 2 musicians and a strange being in the audience.

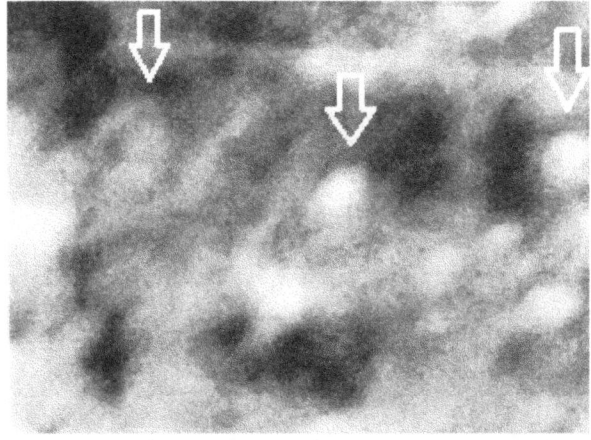

In the image, a man is standing upright and playing guitar on the left. To his left (our right) is a man who seems to be playing a sitar while seated cross legged on the floor.

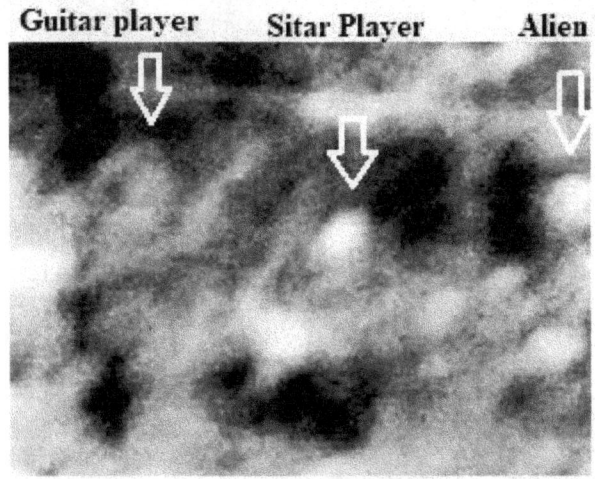

The closer view below reveals that the sitar player has a crown of short black hair and his facial appearance is visible. Also visible are the frets (in white outline) on the neck of his sitar. Important detail again.

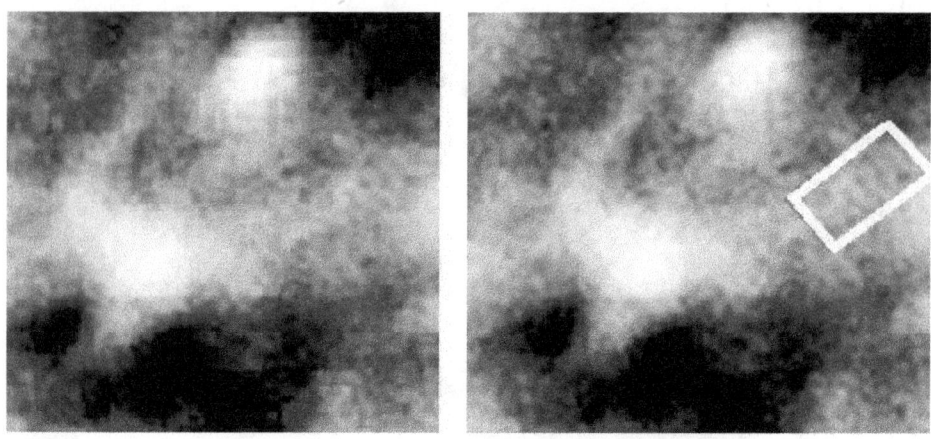

Note what is in the audience to the extreme right of the image. Does this not look like the general conception of an extraterrestrial?

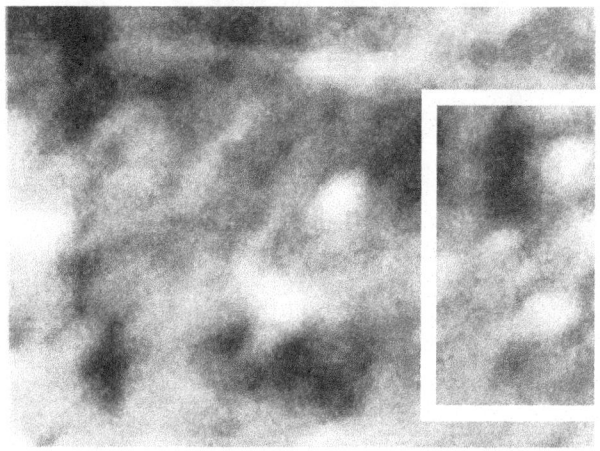

An enlarged closeup of the alien:

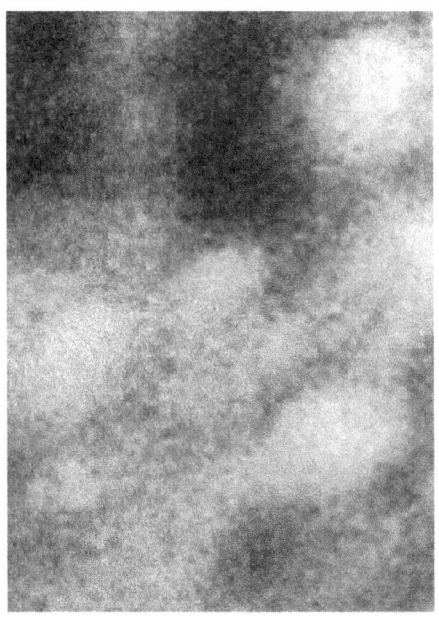

    This other fan from another world seems to have a companion. But his companion, another extraterrestrial, by his different appearance looks like he hails from a different planet. (Notice him below - within the half rectangle). Their nearness of vicinity seems to imply that they are both on Earth together, possibly in the form of 2 separate conquering species (implied by their association together in the picture).

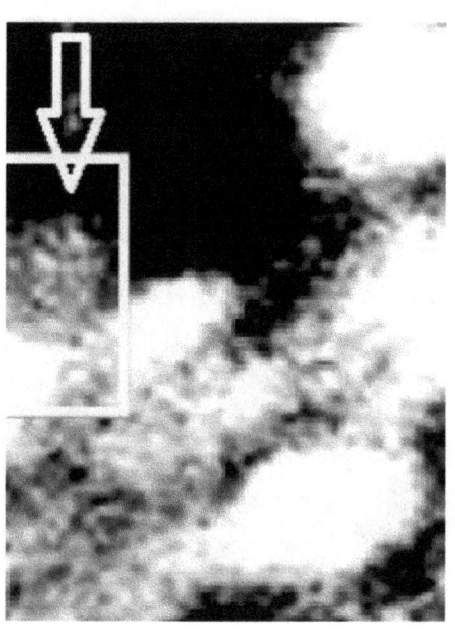

A closer view of the being's face:

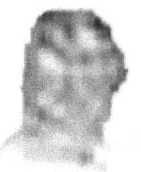

According to these images, a pair of alien beings have come to this location to attend what appears to be a rock concert. And people who have been interviewed about the history of this courthouse have stated that in the 1960's rock concerts were performed here. No one ever mentioned extraterrestrials attending.

But, are these images simultaneous? Is the concert taking place at the same time as the visit from the aliens is occurring or are they overlapping one another in a coded image?

Even the evidence of movement can be verified through the images. Recall the first photograph that was examined in this report (it is from this Bank 2 of images – as all of the ones directly above).

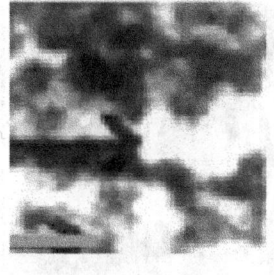

The man in the "polo" shirt was described as sitting in the midst of a crowd. There was another man beside him; the balding man with the van Dyke goatee. These 2 men – and later, a 3rd –

were apparently attending this same concert. Below is the photograph of the original 2 men who were seated side by side in the crowd.

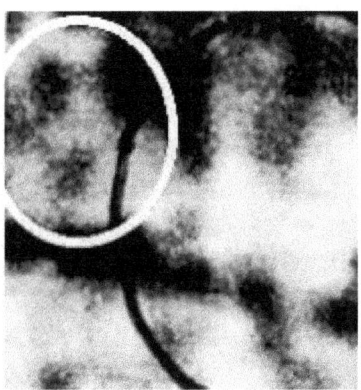

In the next photograph that was taken of this specific location, a 3rd person (?) takes the place of the 2nd one – instantly.

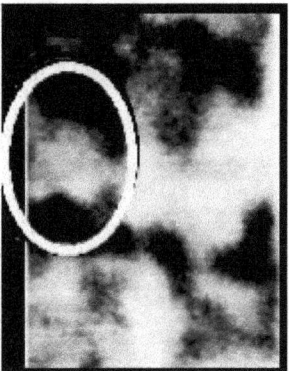

There is suddenly a different 3rd person in the place of the man with the Van Dyke goatee. Instantly because these photographs were taken in rapid succession and the new individual appeared in the very next frame of the photographic record. Person 2 and person 3 are shown below side by side.

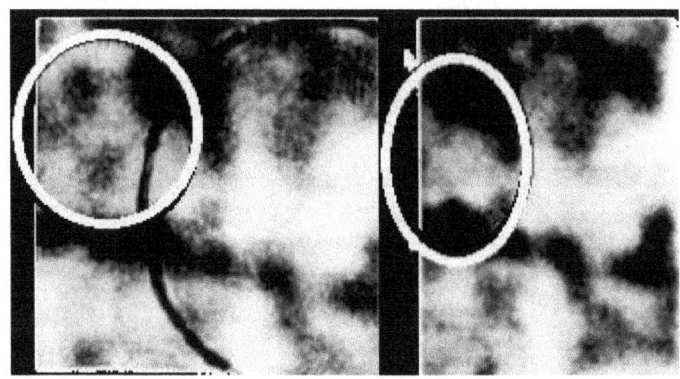

It wasn't an accident that this movement was captured. Many of the photographic sessions were performed in what is called burst mode. This means that a long series of pictures were taken in a steady sequence – one after the other, after the other, etc.. In this way, movement can even be captured by still photography.

This coded messaging shows that movement takes place in this group of images. It also will imply extraterrestrial control of the immediate environment. This is a reason why aliens are probably not the source of these images of warning. Would they warn us of their own takeover?

The upcoming couple of images proves without question that motion was involved in the environment.

The first is a close up of the primary alien's head (Picture A).

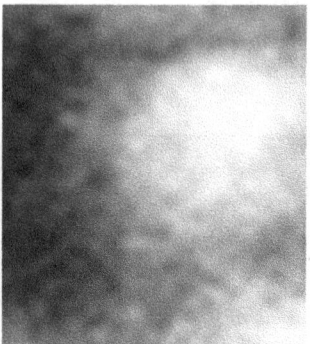

Above is picture A taken of this extraterrestrial. Below is picture B of the same alien:

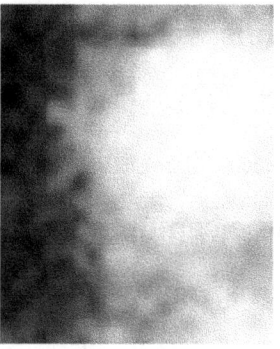

The difference between these 2 photographs is critical! It proves that movement has occurred in the environment in which this "being" is existing! In the first photograph, he is seen in a limited profile. In the second photograph he is seen in a limited frontal face. How could this change have happened unless the subject in the photographed had moved?

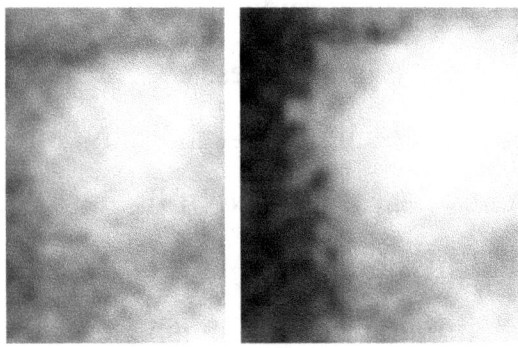

Enhanced:

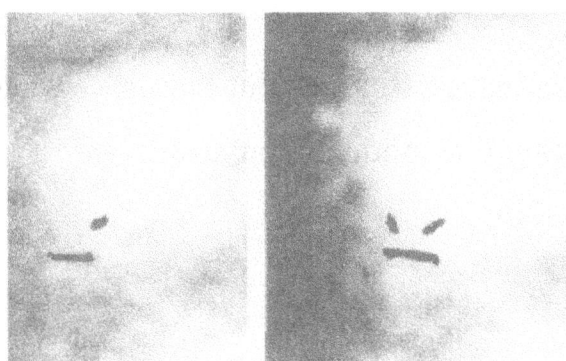

Why is the idea of movement so important? Because it demonstrates that there is an ongoing world energizing these images. Unfortunately, capturing examples of motion was exceptionally difficult. Other examples do exist, but they take place so immensely slowly that it would be almost impossible to translate the tiny about of movement into a photo in a book. This is because of their incredibly decelerated rate of speed.

The imperceptible slowness occurring in the images is despite – and directly counter to – their hypothesized method of transmission, which is near the speed of light. Additionally, it is this hyper speed of transmission which is probably why in the original processing all of the images have a bluish/green tint, denoting speed of light travel.

And perhaps the nature of the transmission of these images is precisely why their visible movement is so slow. Any object – be it image or form – that is propelled forward at such speed would produce the appearance of movement that is of an equally opposite nature to the cause of its mode of conveyance.

666 IDENTIFIED

The magnification abilities of the "hyper-scanner" were pressed to their limit to uncover the digital photographic data which seems to have discovered a being identified as 666. This is as clear a connection to the End Times as can exist! The results of the following pages are going to be astounding!!

This next series of images will demonstrate convincingly that they were definitely the subject of visual cryptographic coding. There will be shown a number of levels or layers of images that were produced for the explicit purpose of revealing the identity of the future antichrist as well as the destruction which is the result of the Apocalypse. It cannot be denied that the following series of images are related and tell a specific story!

Some people have told me that I must've been divinely inspired to have been able to raise the image of 666 from the hell of obscurity in the background of the raw photograph. My view on the subject is that I am a scientist who when a mystery is found will perform all due investigation into it as would any other scientist until either a solution is found or a direction can be found toward that solution. But...something did inspire me.

Examine the next series of photographs and make your own decision.

It begins with an image already examined. This is from Bank 1, which is crowded with cryptographic images (and which are in motion).

This is the chaotic scene which up until this point includes only the 2 persons wearing hazard suits, the crying penitents, and the Orthodox bishop.

Hazard Suits

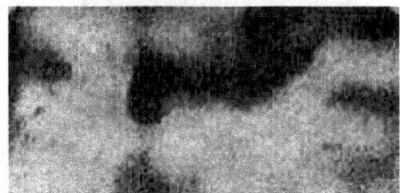

And the one woman in tears:

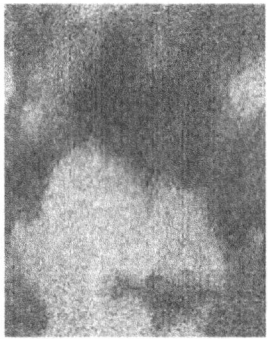

After performing the primary evaluation of this full image – the results of which were produced earlier in this report – another matter concerning this particular image grasped my attention. Focus was placed on a tiny spot on one of the hazmat uniforms – literally a spot. Would it be possible to magnify it to such a degree that its identity – even if just a smudge of dirt – could be determined? That was the thought.

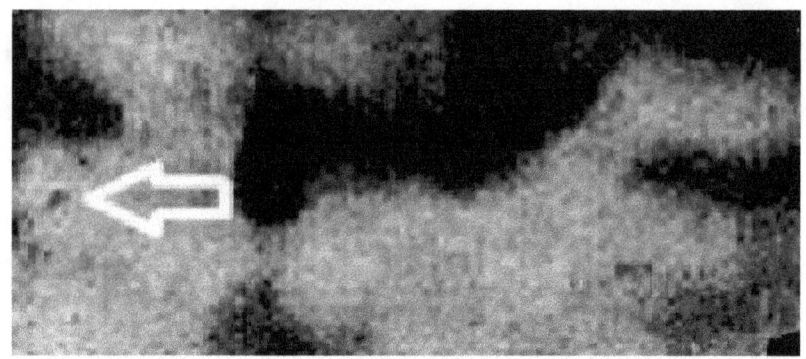

The tiny area toward which the arrow is pointing in the mid-left side of the photograph is what was targeted for extreme magnification. Even this view is enlarged.

Normally, when magnification as extreme as needed for this examination is applied the result is an explosion of pixels.

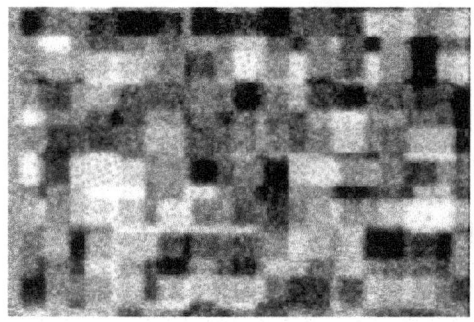

But, in this case, the following is what was produced.

Enlarged it looks like this. A pair of faces. One atop the other.

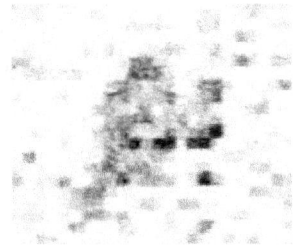

Focus for now only on the top, primary face. 666.

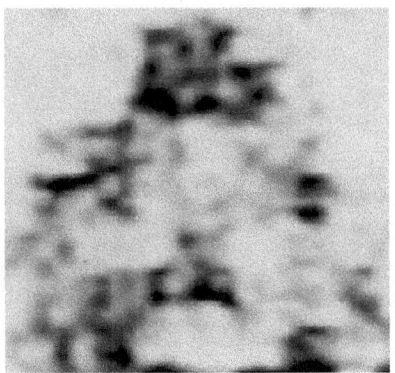

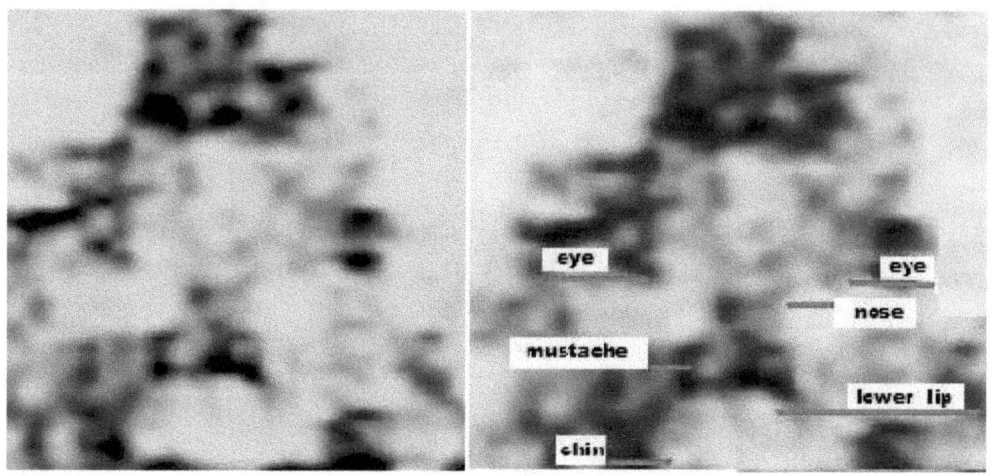

And on his head he wears some form of hat. But who is that who seems to be attached to his lower chin? In biblical terms, he could be described as the prophet of the antichrist's, the one assisting him in being accepted by the people. Due to the positioning of the faces, it is implied that the topmost face belongs to 666 and the one beneath it is subordinate.

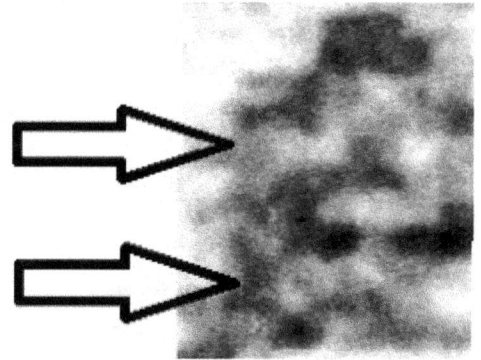

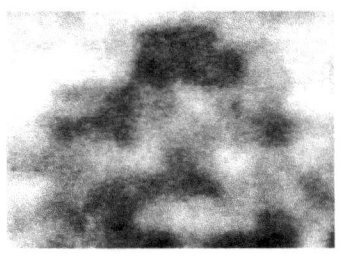

666

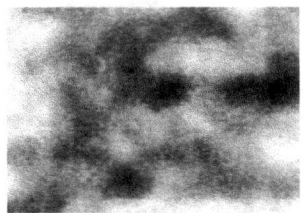

Prophet

But there is something that leaves a person breathless with wonder about this specific picture. And it points directly to proof that this is a creation of visual cryptography by sources unknown!

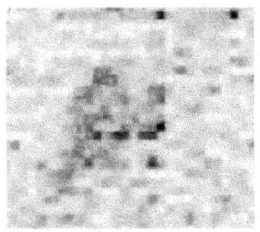

Looking very closely at the above image, one can notice that the 2 faces seems to be contained, or outlined, in an almost imperceptible white frame. It is shown more clearly in the following image, using a black outline in place of the white outline in the original.

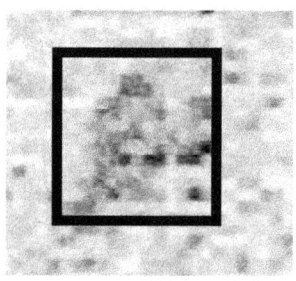

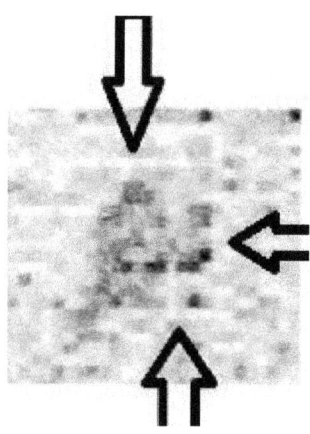

It gives the impression that those 2 faces were cut and pasted into the scene! But by who, when and...HOW! I am not aware of the existence of a photographic process that could operate on such an infinitesimally tiny level to produce such an image! Nor why it would have been inserted into a random photograph taken of a random location. Or why it would be perpetrated – unless maybe as a warning of the coming of 666 into our world and what he will look like.

Aside from God, or the Universal Intelligence, who else could have the immense ability to perform such a feat of metaphysical engineering? Possibly Artificial Intelligence. God or Universal Intelligence seems to be ruled out of the equation because he would not need to resort to cut and paste, just a thought would suffice to create the image desired. But outside sources may have been used.

But what identifies either face as 666, the antichrist, and his prophet? What was found next to them in the photograph below provides one or the other's identity.

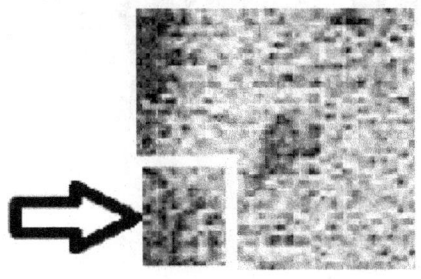

Within that small white square at the bottom left edge of the frame is a series of numbers all of which are arranged in groups of three 666's. It was exceptionally difficult to see these numbers in regular form, so a negative version of the same above picture was created.

The groupings of 6's are circled and the negative image face that also resulted is pointed out by the arrow.

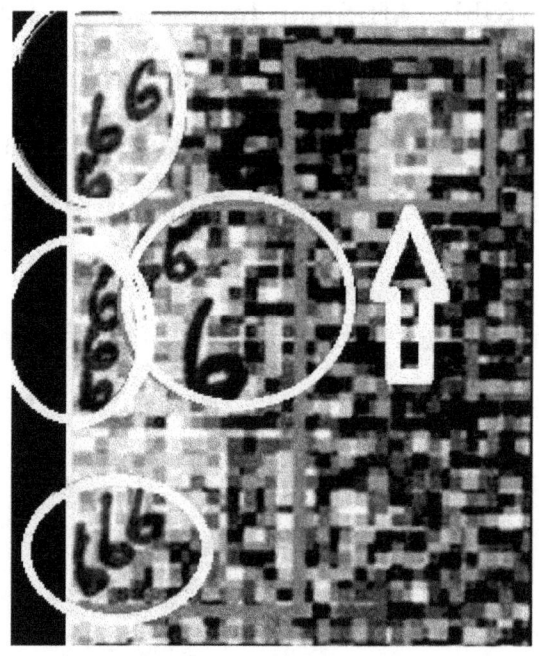

It isn't known how the numbers came to be here, just that they are. And they certainly point to a method of identifying one of the two faces just reviewed. And, as to the source of the warning: it obviously would not be provided by the future antichrist or his supporters, warning us of their coming.

Another powerful clue: notice what happened to the face of the 666 (top face) when subjected to reverse imaging. The negative version of 666 bears a striking resemblance to the face on the famous Shroud of Turin, charging the antichrist as being of the opposite Nature from the Savior, Jesus. And identifying 666 in that way because the bottom face (prophet) did not appear in negative form.

The magnification of the image of the antichrist will be extended to an even deeper level. In many prophecies of the Last Days when the antichrist rules the Earth it is said that he will rise to total power upon the performance of an astounding miracle. He would be shot in the head with a firearm, die, and then miraculously raise from the dead.

Employing deeper magnification, a photograph showing the antichrist with an apparent bullet hole to the forehead was secured. Note the next images.

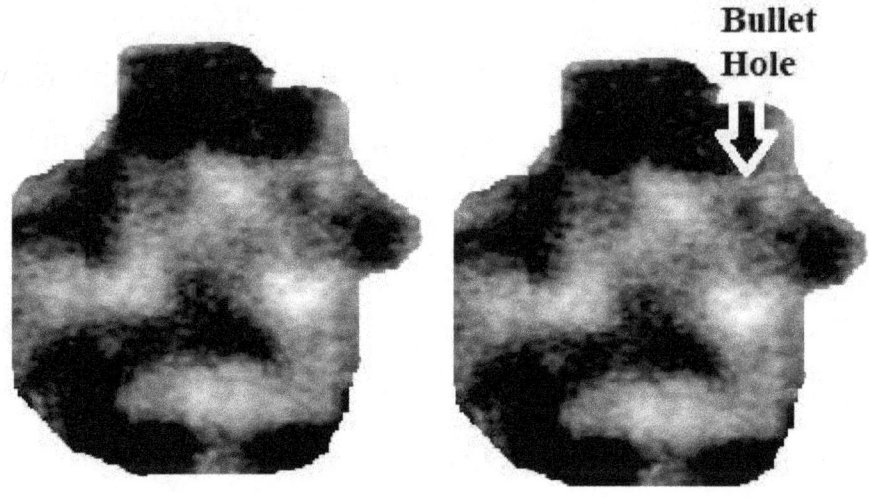

There is yet another level of observation that can be applied to the image of 666 that links it directly to the Apocalypse.

This evidence is found in the background that is in place behind the image of the 666. Thousands, if not millions, of human skulls! When the background is magnified, a massive heap of human skulls with empty eye sockets and disfigured craniums is revealed. Does this represent the worldwide carnage that remained – or will remain – after the Apocalypse? An apocalypse we can avoid?

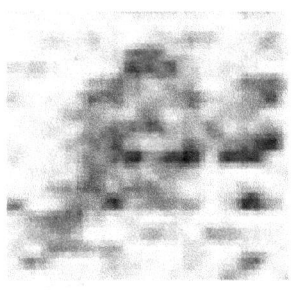

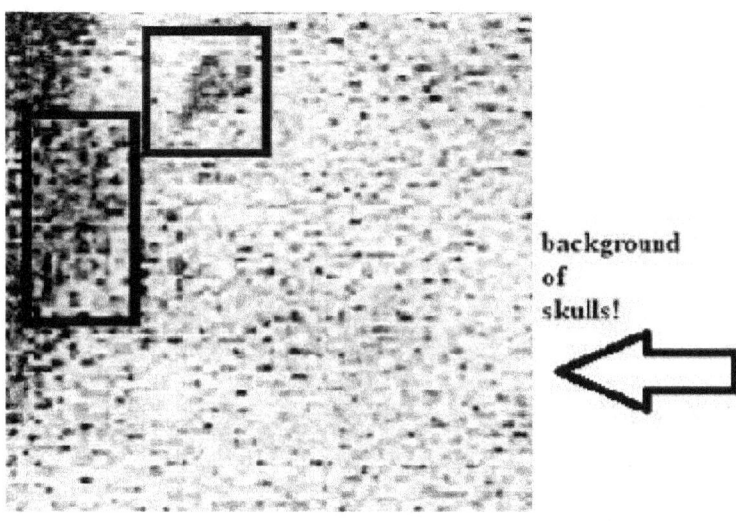

background of skulls!

We have been supplied with the facial features of 666 and have been warned about the murderous results of his rule. Is there a name by which he can be identified?

Found within the same image containing the 2 hazmat suits is a somewhat artistic looking pattern of interlocking symbols, something like a mosaic. It has a hypnotic quality about it.

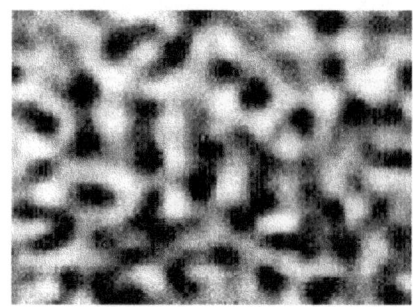

Within this chaotic yet mesmerizing image, a grouping of letters emerged when the proper background is erased. It produced what could be a person's name – the name of the antichrist. Or perhaps another name could be spelled out of a different combination of these letters?

Is this Tivew?:

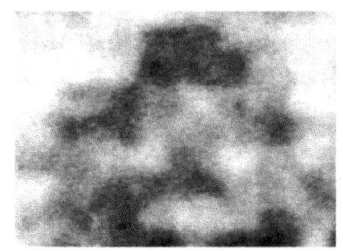

CODES IN LIVING GLASS

Living glass? Can the prophetic window panes in this report feel pain? Certainly an original concept. But when dealing with the phenomena exhibited by the windows in question it has become necessary to develop unique and original nomenclature.

The codes about to be examined are biological of nature and seem to exist in a living medium within the window glass itself. Do they identify a type of bacterium which causes plague? Genetic coding material to be inserted into the normal human DNA to alter the species into some monstrous new creature for the benefit of extraterrestrial or Artificial Intelligence masters?

The next image to be decoded exists in the familiar Bank 1 of images. But it is restricted to another area of the scene, seemingly in some fashion existing in a different type of physical environment and segregated by itself. It's message of warning is different; it involves direct biological manipulation.

Below is the single window pane where the section of "living" glass is located. It's one shown earlier. The "living" glass portion is restricted to the right side of the entire image (within the rectangle).

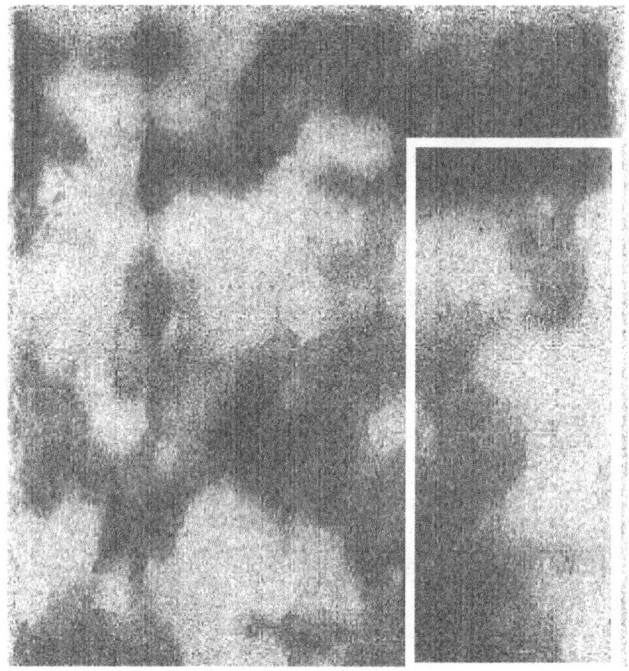

Located within the rectangle is the form of a mostly nude female being who is standing with her back to the viewer. Her purpose and specific identity are unknown other than that she seems to be a "carrier" of, or a victim of experimentation with, biological processes. She seems to be immersed in a living culture – appearing all around her – as if she might be floating in a life size bacteriological petri dish.

The woman is below.

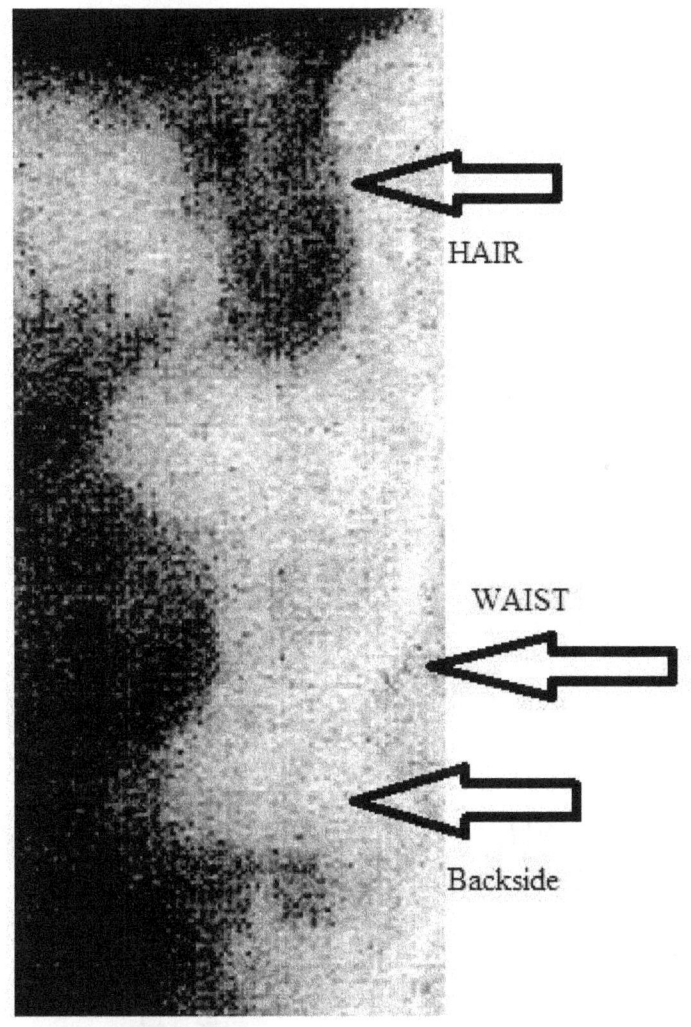

Next is a magnified profile view of the woman's particularly masculine looking face. This seems to imply a loss of gender identity in the planned "new human being."

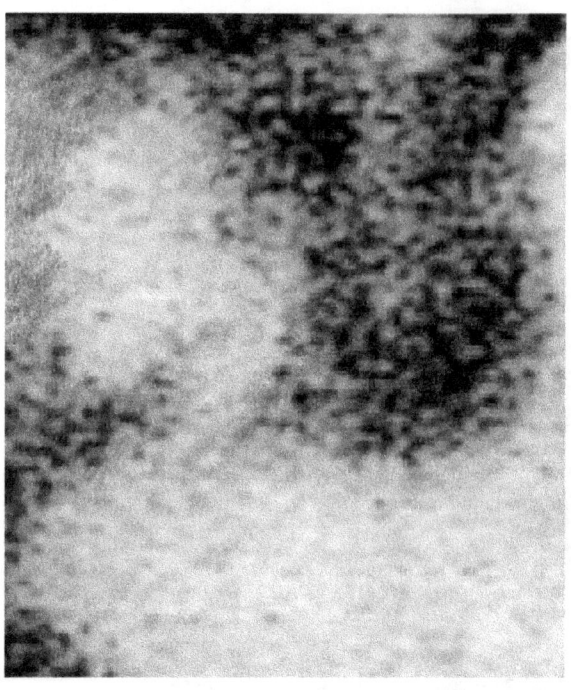

Looking very closely at this picture, it can be seen that she is awash in codes, lettering, and all manner of odd shapes. Some of the coding is much darker than others, as below.

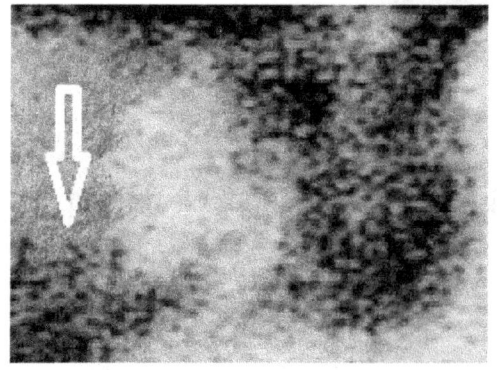

Closer view:

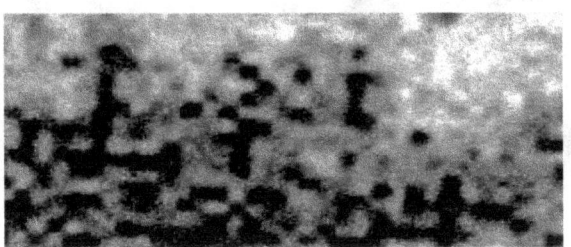

Artist's written copy:

There were also many odd symbols found next to and on her body. As such, the ciphers would have to be floating in mid air or within some form of transparent gelatinous medium of unknown substance.

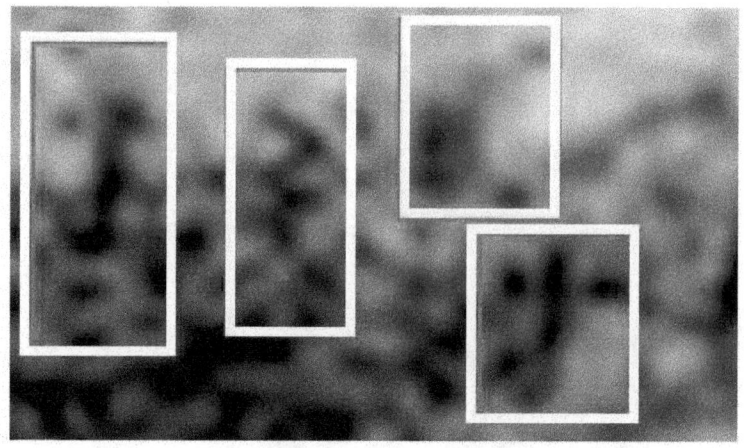

Yet, they seem to have been "stitched" into the sub-level of the image. Evidence of healing implies that these symbols were in someway grown onto the mysterious window medium at this location. They have the appearance of healed over scars. This is seen in the tiny areas of white that cross over the symbols at various points. Healing implies time passage.

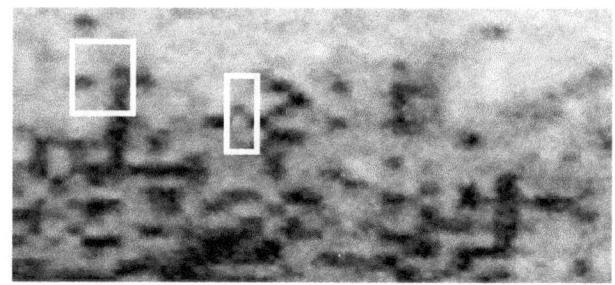

How this scarring occurred is a mystery when it is recalled that the photographs were taken of blank window panes, unless the glass is alive somehow. These symbols could not have been etched onto the panes because there aren't any physical markings on the glass.

When a wider view of the area is taken, the mystery also expands – greatly! Everything seems to be floating in some form of gelatinous medium.

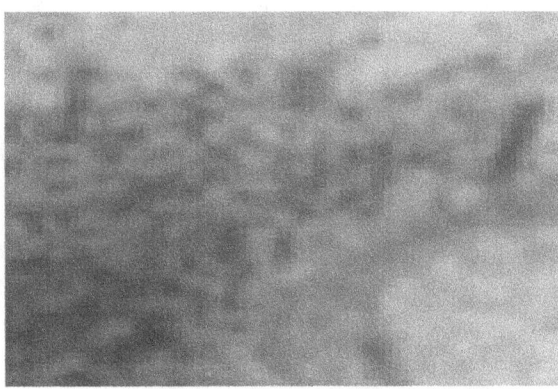

Below is an extreme close up of a portion of the preceding image. Remember, this is in a photo of a blank window!!

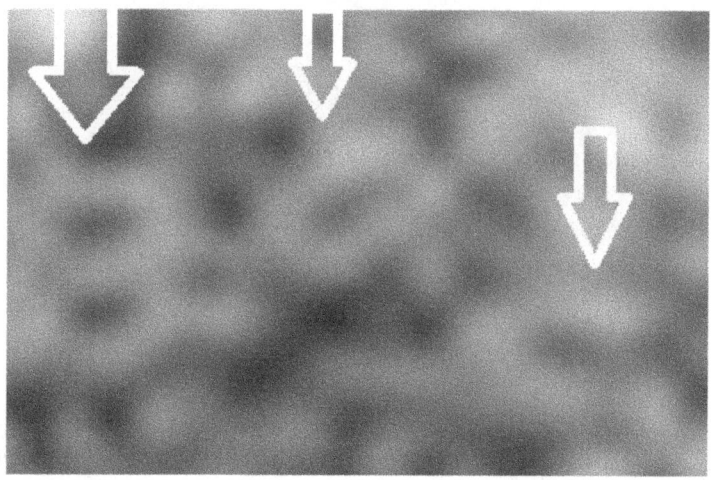

The arrows are pointing at 3 different symbols which are like 3 letters in the English alphabet: B, U, and a sideways Y. That they exist is clear. What they mean is not. Possibly they refer to an alien

DNA code which uses B, U and Y instead of A, C, T and G which are basic to the human DNA code.

When performing a magnification of the above female's lower back portion (see arrow), 2 other types of code were found: the letters F, S, C.

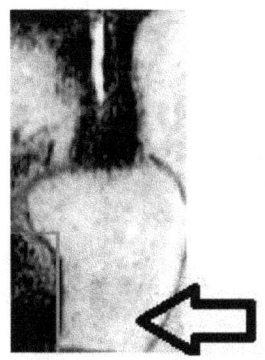

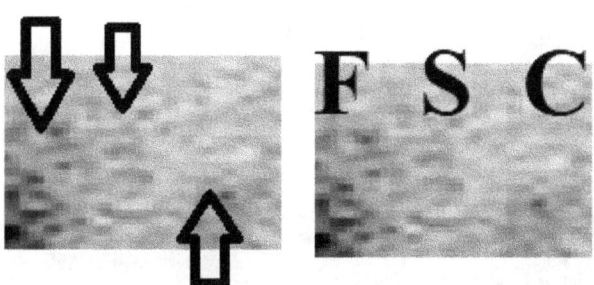

The next code is similar to the type that was used in the early days of computers when punch cards were used. Below is a standard punch card.

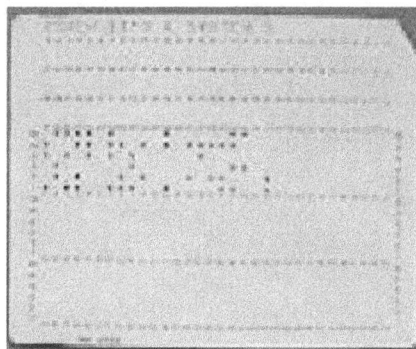

Below are the codes also found in the back portion of the female in the image.

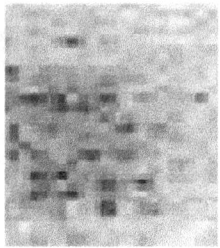

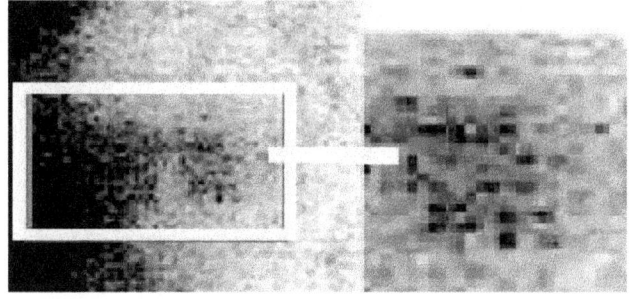

Is there any message that can be surmised from these

"codes?" The basic clue seems to be that they are all centered around a female figure. Females are the ones who bring new life into existence. The symbols seem to point to a new form of DNA. Is the message being given: a new form of humanity will be developed following the Apocalypse at some indefinite time in the future? Maybe the woman of the images is Eve of the Apocalypse? Or Maybe she is going to be infected with a new strain of deformed humanity which will destroy the world. Or a form of plague?

If any of this is true – what are we supposed to do with this knowledge? All of which was obtained in the medium of "living" window glass which was made accessible to us by a cosmic strike made by a particle-sized black hole filament in 1858? Science-fiction? But, these are the facts.

DEMON AND FRIEND

One item appeared in the following window pane in Bank 1 in the pane on the bottom right.

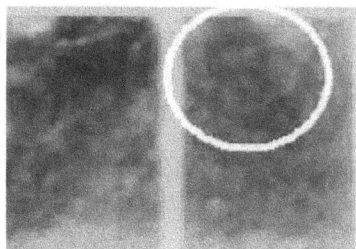

It is difficult to analyze since it seems to be self explanatory, yet the explanation is open to many interpretations. The image in the circle appears to be that of an aged, demonic creature, whispering into the ear of a young man or older boy.

This is obviously of exceptionally poor quality and is only added here as a curiosity; although there is definitely an image present. For those of you who still cannot define what's in the image, a quite accurate depiction arises when certain elements are lightened and others darkened. Nothing is added or subtracted.

And each individual pictured:

Older Boy or Young man

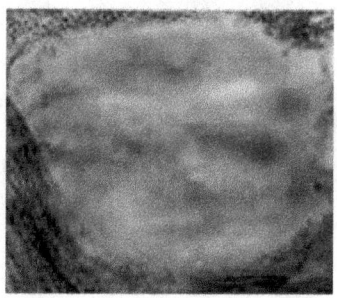

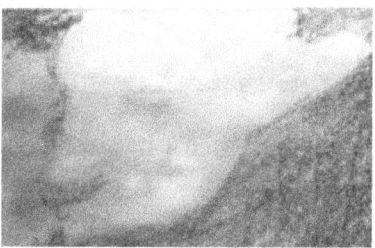

Aged demon

The above should be compared to a similar image of uncertain messaging below. The same window, top right pane.

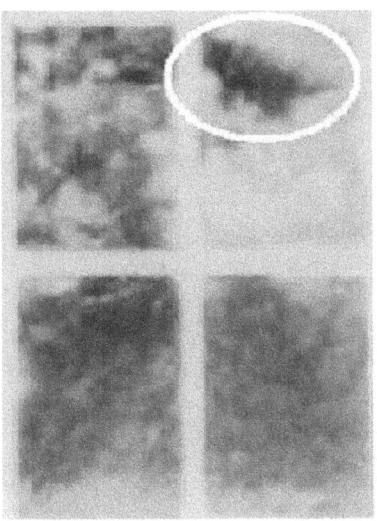

A black cat is in the pane just above the one bearing the young man and the demon. It's appearance is much sharper, but its purpose is as confusing.

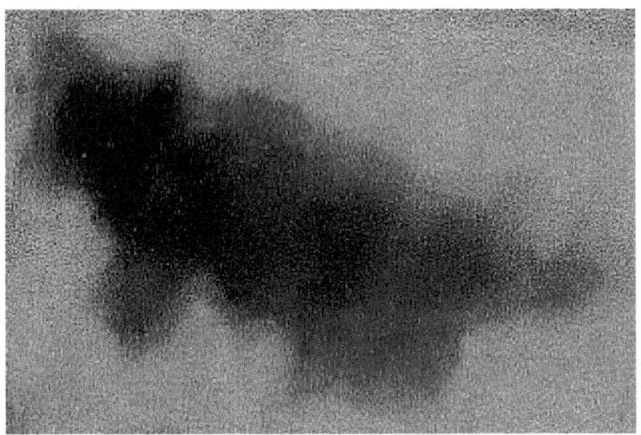

What this pair of images have in common is that it occur in separate panes, they are the only images in their respective panes, and they are not associated with any form of coding. In their case, the source would probably be thought forms.

The black cat may have frequently been in the thoughts of someone who'd worked in this courthouse and, so intent were the thoughts, that they became etched into the window glass. Window glass by the way which had retained those special qualities imbued by the mysterious "windstorm" of 1858.

The young man and the demon probably also owe their source to thought form creation, but their reason for creation is more difficult to conjecture.

Although most of the images in the window panes are of a prophetic nature that does not mean that all of the images have to abide by this restriction. As noted above, the special magnetized powers with which some of the glass had been saturated could have different reactions to different sources, producing different effects.

ABSOLUTE PROOF

While originally investigating this story, interviews were made with all of the primary persons involved with the building in question. Among this group were: the building engineer, the county archivist, members of the city council in charge of the building, as well as the day-to-day public workers involved with maintenance of the structure.

Many of these individuals directed me toward one person in particular, Mary Hart, whom it was claimed was the ultimate expert on the building in question. However, this Mary Hart was never available. She had either just gone out of town on business or to visit a family member, or had just left for the day, or had taken the day off. I never got to interview Mary Hart. Probably because Mary Hart never existed.

I have been employed as a journalist for several years, under a different name than that used for my books. I even had a couple of bylines placed with the *New York Times*. A reporter quickly learns when he is being given, in colloquial terms, the run around. And the non-existent Mary Hart type of obfuscation usually denotes a widespread cover-up.

The many interviews that were made may have been a mistake in one sense. It caused those in power to know that an investigation was taking place on a matter that they wished to keep secret. And, eventually, they acted and made any further investigation impossible. But by doing so it provided incontrovertible, undeniable evidence that the phenomena that had been under investigation was genuine.

Since the finding had been made about this unusual building, I maintained personal surveillance over it. I often returned to take more photographs in search of more evidence.

But, one night in 2011, when the images of the primary windows that had been taken that day were subjected to computer processing I received an alarming shock. The windows at the center of the investigation had been altered. They weren't replaced: they were altered – changed. Somehow, the people in charge of the structure had found a way to "cleanse" the windows of any any digital data – the images – that at one time had been electromagnetically impressed into them.

The pictures below compare the primary windows that once held the images. The window on the left is how it looked in 2008. The window on the right is how it looked after it was altered. They look the same, and on the "surface" they are – except for the putty.

2008   **2011**

Below is a closeup of the 2011 window showing the putty smudges that were carelessly left by the crew after removing, then replacing the SAME window glass.

This author can verify it is the same glass because on close examination it proved to be the same. Remember, I also have a background as an industrial engineer, specializing in glass production and design. Thus, some form of technical electromagnetic cleansing operation must have been performed on the glass. Below are the images of negative proof.

It is basically a before and after view. **Before** is the way the windows looked while in the original processed state. **After** is how they appeared after processing in their new "cleansed" condition. LEFT: 2008. RIGHT: 2011(white spots – putty).

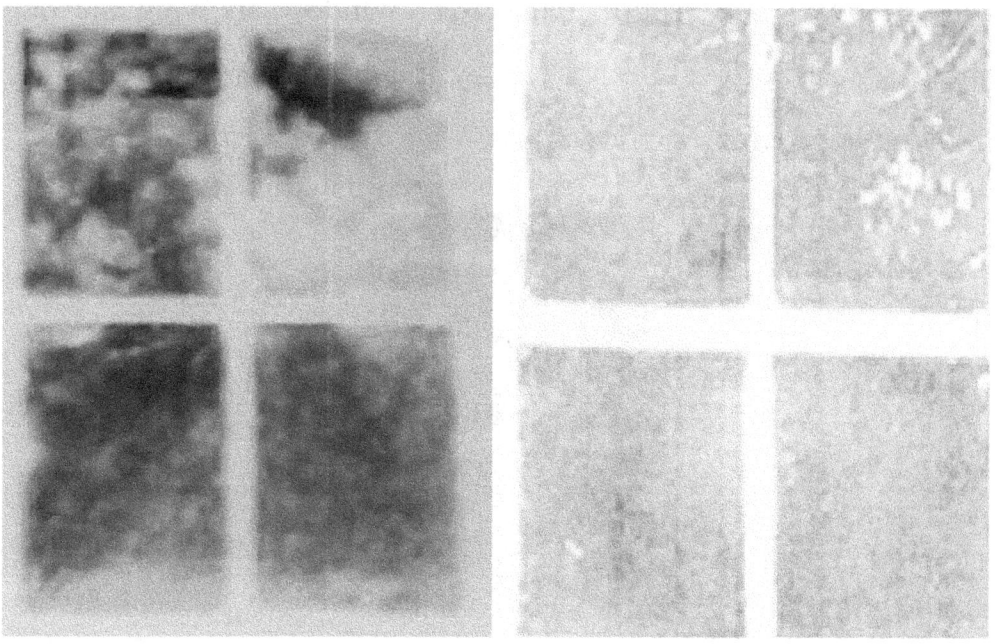

Some agency of immense power seems to be guarding this building and attempting to prevent any knowledge of its secrets being made known.

But there is even more! The destruction of the hyper-scanner that allowed for the deepest magnification of the images. After having completed this book and others that disclosed NASA censorship of exploration of the Moon and Mars, this author returned to his former home to find the "hyper scanner" vandalized so it could never be used again. Nothing else was harmed and nothing was stolen. The only target was the scanner and – the truth!

FINDINGS

A mass of data was collected during this investigation. Most of the evidence resides in the images taken and the method used to acquire them: a photographic method known as quantum imaging.

Proof that the phenomena was genuine was supplied in negative ways by the individuals who oversee the building in question. The sudden evacuation. The lock down of the site. The failure to divulge the specific reason for these actions. And finally – and most damaging – the removal of the very physical evidence that they refused to admit even existed. If there was nothing to hide, then why hide it?

It is to be hoped that the data that had been transmitted onto the window glass from sources yet unknown has been retained by whatever power removed it and that the information is being well preserved. At any rate, the original data is still in my possession and safely secluded.

It is regrettable that data of a more scientific and physical nature about the building and the images could not have been secured: radiation level readings, sonic and other frequency types recorded, and similar testing procedures undertaken. This was not allowed since entrance to the building was forbidden and attempts to obtain this type of data from the exterior through the sturdy walls was not possible.

This investigation began in 2008. The findings are finally published in 2020. That involves 12 years of research – at times interrupted.

At the outset, this researcher was confronted by a daunting paradox. Images without apparent source appeared in a few select window panes in photographs taken of a historic building, but only became visible after computer processing which did not involve any alteration to the originals. How was this possible? It isn't. When the only answer to a scientific problem is the impossible, one looks toward quantum physics – or a Higher Intelligence – for a hopeful solution where the impossible is often the correct answer.

The only form of photography that could literally make the invisible visible was quantum imaging. Quantum imaging is an agonizingly complex process and it required months of study to assimilate the barest understanding of the principles by which it operates. This took time.

What also took time was testing hundreds – probably thousands – of randomly chosen windows from random locations that were photographed with the same process as used for the mysterious windows in the historic courthouse. Did any of them reveal similar images? After an exhaustive, time consuming study, not one image was found. Anywhere!

The uniqueness of the images discovered in the special window panes was established. They were real and a rarity. It is my belief that some of the images are from other time periods that existed within this building and that they are still ongoing and have been revealed by the effects of a particle-sized black hole filament strike in the past.

More directly, however, a warning has been transmitted by an unknown source about our potential future, possibly based on transferring images to our realm from a duplicate history that is occurring in a parallel universe to our own. The warning is: we are facing an imminent apocalypse over which an antichrist will rule under the control of alien masters, and that during this period the human species will be re-created by the imposition of a new genetic code. Ultimately, a "race" of Artificial Intelligence will rise to power. The source of this warning is still unknown.

And, to conclude on a light note: still seeking the owner of this cat:

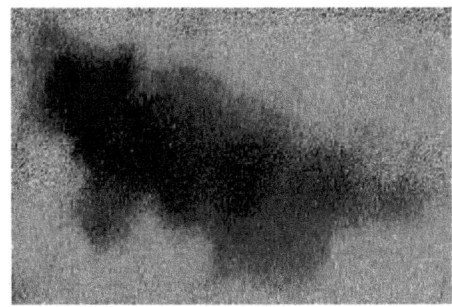

The End

February 29, 2020

www.ingramcontent.com/pod-product-compliance
Lightning Source LLC
Chambersburg PA
CBHW080252030426
42334CB00023BA/2794